THE EMBODIED MIND

THE EMBODIED MIND

UNDERSTANDING THE MYSTERIES OF CELLULAR MEMORY, CONSCIOUSNESS, AND OUR BODIES

THOMAS R. VERNY, M.D.

PEGASUS BOOKS
NEW YORK LONDON

THE EMBODIED MIND

Pegasus Books, Ltd.
148 West 37th Street, 13th Floor
New York, NY 10018

First Pegasus Books paperback edition June 2023
First Pegasus Books cloth edition October 2021

Interior design by Maria Fernandez

Library of Congress Cataloging-in-Publication Data is available.

ISBN: 978-1-63936-462-6

10 9 8 7 6 5 4 3 2 1

Printed in the United States of America
Distributed by Simon & Schuster
www.pegasusbooks.com

To Sandra, my muse and my love

CONTENTS

If we value the pursuit of knowledge, we must be free to follow wherever that search may lead us. The free mind is not a barking dog, to be tethered on a ten-foot chain.

—Adlai E. Stevenson II

INTRODUCTION

As a thirteen-year-old boy I read Sigmund Freud's *The Interpretation of Dreams* in the original German in Vienna. I was totally fascinated by how Freud's slow, methodical questioning eventually led to the discovery of deeply hidden unconscious conflicts in the lives of his patients. Then and there I resolved to become a psychiatrist.

Years later, when I had become a psychiatrist, I continued to be fascinated by dreams and the unconscious. One day, while working with a young man on his dream he suddenly, without any input from me, started to cry like a little baby. He cried for close to ten minutes and then stopped on his own. What just happened? I asked him. He told me that in his mind he found himself in a crib and that he was crying for his mother. Then, he recalled that he had actually seen photos of himself as an infant and some of them pictured him lying in a blue crib whereas the crib that he had just experienced was definitely white. He wondered about the discrepancy.

I suggested that he ask his mother to resolve this question. The next week he returned for his regular appointment and told me that according to his mother, when he was born his parents lacked money for a new crib but were able to borrow one from a neighbor. The borrowed crib was white. A few months later, they were able to buy a new crib for him and that new crib was blue. That is the one of which all the photos were taken.

I felt both intrigued and mystified by this experience, since throughout my studies first at the University of Toronto then Harvard University, I was taught that children remember nothing before the age of two. And yet as I continued to practice, I repeatedly encountered patients who would tell me about events in their lives reaching far back in time to infancy, birth, and even womb life. A few of these memories may have originated from overheard conversations by family members or gleaned from photo albums or videos. However, a considerable number would not have been easily available and were corroborated by evidence supplied by parents, hospital reports, and other documentation. I wondered how to explain these memories scientifically. It was then that after much study, research, and personal contacts with colleagues in obstetrics, psychology, psychiatry, and other sciences, I wrote *The Secret Life of the Unborn Child*, which is now published in twenty-seven countries and continues to enjoy wide popularity.

At the time, almost forty years ago now, I had much solid scientific evidence to back up the central premise of my book; namely, that an unborn child is a sensing, feeling, conscious, and remembering being, at least three months before birth. However, I had little or no scientific evidence to support cognition of any kind extending back further in time. Of course, given the rapidity of development and change in the biomedical sciences these past decades, forty years is practically an eon ago. Much of what is now known in cell biology, genetics, and more important, epigenetics, not only confirms my claims in *The Secret Life*, but enables me to put forward the bold new concepts in *The Embodied Mind*.

What set me on the path toward *The Embodied Mind* was an article I read six years ago reprinted from *Reuters Science News* titled "Tiny Brain No Obstacle to French Civil Servant." It seems that in July 2007, a forty-four-year-old French man went to a hospital complaining of a mild weakness in his left leg. When doctors learned that the man had a spinal shunt removed when he was fourteen, they performed numerous scans of his head. What they discovered was a huge fluid-filled chamber occupying most of the space in his skull, leaving little more than a thin sheet of actual brain tissue. It was a case of hydrocephalus, literally—water on the brain. Dr. Lionel Feuillet of Hôpital de la Timone in Marseille was quoted as saying, "The images were most unusual . . . the brain was virtually absent." The patient was a married father of two children, and worked as a civil servant apparently leading a normal life, despite having a cranium filled with spinal fluid and very little brain tissue.

To my surprise, I found in the medical literature an astonishing number of documented cases of adults who as children had parts of their brain removed to heal their persistent epilepsy. Following hemispherectomy most children showed not only an improvement in their intellectual capacity and sociability but also apparent retention of memory, personality, and sense of humor. Similarly, adults who have had hemispherectomies enjoyed excellent long-term seizure control and increased postoperative employability.

If people who lack a large part of their brain can function normally, or even relatively normally, then there must exist, I thought, some kind of a backup system that can kick in when the primary system crashes. I devoted the next six years to studying the medical and scientific literature, searching for evidence to support my hunch.

I found that while many scholars had contributed greatly to advancing science in their own fields, no one had really synthesized this knowledge, "connected the dots," and thought of addressing this puzzle. *The Embodied Mind* attempts to do just that.

Our embodied mind is not the old enskulled one. It is an extended mind that relies on the intelligence of all the cells in our body that contain specific bits of information, micro-memories. All memories, consciousness, and the mind emerge from this linked sentient network.

The Embodied Mind, which I will seek to establish as its own unique psychobiological term, represents a coherent and empirically grounded biological theory that marks a significant departure from the past century's exclusive focus on cortical neurons (brain cells) as the only important cells in information processing, cognition, and memory storage. *The Embodied Mind* is based on studies that demonstrate intelligence and memory in a wide range of systems well beyond the traditional central nervous system, including the immune system, sperm and ova, unicellular organisms, amoebae, and many more. Memory is truly a body-wide web. Whether or not we can consciously access a memory is not as important as the realization that we had the experience, the lived event, which has left some kind of impact, influence, mark, trace, record, or imprint on our cells and tissues.

A large number of these effects may be passed on to our children and grandchildren. Therefore, it is imperative that we become aware of as many of our basic maladaptive urges and behaviors as possible and consciously try to overcome them. At the same time, it is imperative for our sake but especially for the benefit of our future children to live a good and healthy life. We shall vastly improve our lives and the lives of future generations by actively avoiding stress and anxiety as well as people who are critical or deceitful and instead befriend people who support and value us.

Like musicians in an orchestra playing in different sections, be they strings, woodwinds, brass, or percussion, the cells in the skin contain different information from the cells in the heart and so on. The memory that emerges either consciously or unconsciously as a result of some trigger is "heard" like the music emanating from an orchestra. The higher brain centers take the place of the conductor and coordinate the messages that reach our conscious self and lead to cognition and behavior.

It is time we put to rest the myth of the enskulled brain and mind and adopted the scientifically evidence-based concept of the embodied brain and mind. This is a transformative, novel concept in psychobiology, at once paradigm-shifting and empowering.

We think, feel, and act with our body. We relate to the world with our body. Our mind is body bound. It is my hope that *The Embodied Mind* will help us gain more insights into who we are in relationship to ourselves, our loved ones, society, and the universe. It will motivate us to exercise our free will and encourage us to take responsibility for our own actions.

CHAPTER ONE

DO GENES MATTER?

Introduction

The union of sperm and egg at conception leads to the formation of a fertilized ovum, a one-celled organism, the zygote, that, if successfully implanted into its mother's womb, will eventually become an adult person. This tiny cell will carry the blueprint for the future of an entire human being. Astonishing but true. What is even more amazing is that on the basis of solid scientific evidence I can say that this genetic information is not limited to just architectural plans for building a body but may also include data reflecting experiences and personality characteristics of the parents. Such *acquired* characteristics catapult us into the new science of epigenetics.

I think it is fair to say that epigenetics is the most revolutionary advance in the biological sciences since Charles Darwin's *On the Origin of Species* was published in 1859. Epigenetics is the study of the molecular mechanisms by which the environment regulates gene activity. Epigenetics teaches us that life experiences not only change us but that these changes may be passed on to our children and grandchildren down through many generations. This process is called *trans-generational inheritance*, and has become a hotly debated area of research.

From an evolutionary perspective it makes good sense that exposure of parents to significant environmental conditions such as hunger, warfare, anxiety, and the like should "inform" their offspring in order to better prepare them to meet these conditions when they are born. Obviously, this information can only be conveyed from parents to their children by way of their germ cells (ova and sperm).

In the last decade, genetic research has established that the DNA blueprints passed down through genes are not set in stone at birth. **Genes are not destiny**. Environmental influences, including nutrition, stress, and emotions, can modify the expression (whether they are turned on or off) of those genes without changing the genes themselves.

We shall take a whirlwind tour through genetics: chromosomes, genes, DNA, RNA, etc. Then we shall move on to epigenetics, which I have divided into environmental epigenetics, which deals with physical environmental factors such as pollution, toxins, too much or too little food, and psycho-social epigenetics, which is concerned with relationships, particularly parent-child relationships, and psychological factors such as stress, anxiety, or the presence or absence of affection. We shall pay particular attention to the impact of abusive and neglectful caregiving and parental adversity on a child's epigenome.

Genetics

It is impossible to discuss genetics without the use of scientific jargon. For this, I ask your indulgence and patience. Even if you find some of these terms daunting, please read on. You will get the gist of it. I promise.

The basic unit of inheritance is the chromosome. A chromosome is an organized package of DNA (deoxyribonucleic acid) found in the nucleus of every cell. Different organisms have different numbers of chromosomes. Humans have twenty-three pairs of chromosomes. These consist of twenty-two pairs of numbered chromosomes, called autosomes, and one pair of sex chromosomes, *x* and *y*. If you have *xx*, you become female; *xy*—male.

Each child receives half of their chromosomes from their mother and half from their father.

A genome is the complete set of DNA in a cell. The twenty-five thousand to thirty-five thousand genes on the human genome make up only 5 percent of the entire genome. The rest consists of switches and long stretches of noncoding DNA (meaning they do not make proteins). These regions between genes were for a long time dismissed as "junk DNA." Scientists have recently learned that the regions between the genes are the switches that play a vital role in cell functions. Mutations in those DNA regions can severely impact our health.

Robert Sapolsky, professor of biology, neuroscience, and neurosurgery at Stanford University, when discussing the human genome says in his wonderful book, *Why Zebras Don't Get Ulcers*, "It is like you have a 100-page book, and 95 pages are instructions on how to read the other 5 pages."

In his seminal *On the Origin of Species*, Darwin wrote that evolutionary changes take place over many generations and through millions of years of natural selection. Following in Darwin's footsteps, geneticists have had remarkable success in identifying individual genes with variations that lead to simple Mendelian traits and diseases (see endnotes) such as phenylketonuria (PKU), sickle-cell anemia, Tay-Sachs disease, and cystic fibrosis. However, diseases with simple Mendelian patterns of inheritance are rare, while most human diseases such as cancer, diabetes, schizophrenia, and alcohol dependence, or personality traits and behavior, are the result of a multitude of genetic and psycho-socio-economic-cultural elements and therefore, considered complex and multifactorial.

Time magazine's covers often reflect a dominant cultural, political, or scientific phenomenon. The October 25, 2004, cover portrayed a woman praying with the inscription THE GOD GENE. It refers to an article in that issue that hypothesizes on the presence of a "God Gene" in our genome. Of course, nothing could be further from the truth.

There is no God Gene, or Anger Gene, or Selfishness Gene, or Schizophrenia Gene. It takes many genes to develop a disease or bring about a personality trait. By the same token, a different combination of the same genes can create high intelligence, musical abilities, foresight, etc. Researchers from the University of Geneva report that genetic variation at a single genomic position impacts multiple, separate genes. **If one element changes, the whole system changes. Genes teach us a crucial life lesson:** *Everything is connected.*

A case in point is the finding that personality changes can affect body shape and body movements, at least in zebrafish (FIG 1.1), as a powerful new study from North Carolina State University demonstrated recently. The researchers bred one group of fish to be bolder and another group to be shy. Zebrafish that were bred to be bold displayed a sleeker body shape and an ability to dart around the water more quickly when startled than those bred to be shy. This study supports the assumption that traits like personality or temperament may be genetically correlated with other traits, like body shape. The body is one complex ecosystem where if even the smallest part changes, everything changes, like the proverbial "domino effect."

The genome's functioning is dependent on its intracellular environment (the environment in the cell surrounding the nucleus) and its relationship to the extracellular environment, including hormones and neurotransmitters. The extracellular environment, in other words, the tissues and organs of the body outside the cell, are in turn affected by the environment of the individual—for example, by the availability of food or social interactions. Consequently, we unconsciously adjust our lives to everything that transpires inside and outside of us. It's wonderful that our body can do this on its own. We don't even have to think about it most of the time.

With a few exceptions, every cell type in a multicellular organism carries the same endowment of genetic instructions encoded in its DNA genome. Nevertheless, each cell type expresses (activates) only those genes required for its specific performance of function. The proteins that package the genes in the cell nucleus are called histones. Histones act as spools around which DNA winds (FIG 1.2). Histones play an important role in gene regulation. More on this in the next section.

The dominant view of heredity is that all information passed down from one generation to the next is stored in an organism's DNA. Very recently, cellular biologist Antony Jose has advanced a new, we might say revolutionary, theoretical framework for heredity. Jose challenges the common view of heredity that all information passed down from one generation to the next is stored in an organism's DNA and argues that DNA is just the ingredient list, not the set of instructions used to build and maintain a living organism. The instructions, he says, are much more complicated, and they are **stored in the molecules** that regulate a cell's DNA. Jose's new framework recasts heredity as **a complex, networked information system** in which all the regulatory molecules that help the cell to function can constitute a store of hereditary information.

Jose's framework helps us to understand how the storage of information has evolved with complexity over the millennia that must include now the cytoplasm and the cellular membrane in addition to the nucleus. It reemphasizes the need to abandon outmoded concepts of central control mechanism and instead introduce concepts of networks and feedback loops.

The early twentieth century geneticists' view of heredity saw the development of an organism as a one-way flow of information from nuclear DNA to messenger RNA to protein production. This model, also known as *the central dogma of genetics,* is now being superseded by the recent rise of epigenetics. As we shall see, epigenetics is based on the ways that extranuclear factors interact with genes to bring about the changes in an individual.

Epigenetics

Another cover of *Time* magazine, in early January 2010, also depicted a double helix of DNA, this time as a giant zipper hanging down across the cover, its shiny gold slider opening part way, as if unzipping an actual strand of DNA. This time the cover story was: "Why Your DNA Isn't Your Destiny: The new science of epigenetics reveals how the choices you make can change your genes—and those of your kids." This time, *Time* was on the right track.

While Darwin's work defined evolution as a process of incidental, random mutation between generations and survival of the fittest, the new science of epigenetics is much closer to the greatly maligned theory of French biologist Jean-Baptiste Lamarck, who suggested that an organism can pass to its offspring characteristics acquired during its lifetime.

Epigenetics is the study of changes in gene activity that do not alter the genes themselves but still get passed down to at least one successive generation. These patterns of gene expression are governed by the cellular material—the epigenome—that sits on top of the genome, just outside it (hence the prefix *epi*, which means "above"). A key component of epigenetics is methylation, in which a chemical group (*methyl*) attaches to parts of the DNA—a process that acts like a dimmer on gene function in response to physical and psychosocial factors. Epigenetic "switches" turn genes on or off, and all points in between (FIG 1.3, 1.4).

Methylation is a dynamic process, and levels of methylation can change from moment to moment and over the course of a person's lifetime depending on the person's experiences, whether these be external or internal. The opposite process to methylation is acetylation. Methylation turns down or totally silences the function of a gene while acetylation turns on the gene, partially or totally.

It is through epigenetic switches that environmental factors like prenatal nutrition, stress, and postnatal maternal behavior can affect gene expression that is passed from parents to their children. Epigenetic changes represent a biological response to one or more environmental factors. These factors may be positive and life affirming or negative and life threatening. Epigenetic changes serve a very important function during pregnancy by biologically preparing offspring for the environment into which they will be born. **Think of genetics as the hardware and epigenetics as the software in your computer.**

One of the primary objectives of epigenetics is to study data transfer from one generation to the next by biological rather than psychological means. Biological inheritance speaks to the idea that the germ cells (sperm and eggs) are affected by significant environmental events, and that these changes in the genome are then passed on to descendants. Epigenetics offers us the knowledge and the means by which we can enhance physical and mental health, both in our offspring and ourselves.

The union of sperm and egg at conception leads to the formation of a zygote (a fertilized ovum). This tiny cell will carry a set of complete instructions for building an entire human being. I wondered: Is the information limited to just architectural plans for constructing a body, or does it also include data that will affect the mind? Before we move on to address this question, we should mention three other biological ways by which information may be exchanged between people that do not involve germ cells.

It has recently been discovered that some of the cells carried in the blood that pass between mother and child during pregnancy remain in their bodies. Also, a few cells from prior pregnancies persist in mothers for many years. This process is called *microchimerism*. Human and animal studies have found fetal origin cells in the mother's skin, bloodstream, and all major organs, including the heart. What these studies show is that each of us carries two different cell populations, our own plus one from our mother. Women who have carried a child harbor at least three unique cell populations in their bodies—their own, their mother's, and their child's.

Similarly, *blood donations* and *organ transplants* can pass information on a cellular level to a recipient. If my hypothesis of cellular memory is correct, then these "donor" cells may, as in the case of microchimerism, affect their recipients' minds and bodies in ways we are just beginning to explore.

Environmental Epigenetics

In this section we shall discuss how physical factors such as food, nicotine, or odors affect the genome.

A 1988 paper published by John Cairns in *Nature*, one of the most distinguished science journals, started a tectonic shift in genetics. The paper described an experiment in which a particular strain of bacteria, E. coli, that could not metabolize lactose (a sugar found in dairy products), was placed on a lactose medium (scientific jargon for food on which bacteria grow, usually in a petri dish). Instead of starving—which according to classical Darwinian theory they should have—the bacteria very quickly underwent genetic changes, allowing them to digest lactose and thus survive. Cairns reported that at least in some cases, *selective pressures* could specifically direct mutations. Good-bye Darwinist orthodoxy.

Cairns "brazenly," as some critics said, raised the specter of possible Lamarckian hereditary mechanisms—one could not have been more heretical than that in 1988. In the same issue of *Nature*, Franklin Stahl, emeritus professor of biology at the University of Oregon, endorsed Cairns's conclusions and presented his own model of how "*directed mutations*" may take place.

Cairns today is professor of microbiology at the Radcliffe Infirmary, Oxford University, and remains a recognized leading authority in mutation genetics. His 1988 article is one of the most frequently cited papers in the field, and has launched an entire new area of study.

At about the same time as Cairns was performing his experiments, Dr. Lars Olov Bygren, at the University of Umeå, Sweden, wondered, "Could parents' experiences early in their lives somehow change the traits they passed to their offspring?" Bygren and many other scientists have now amassed abundant historical evidence suggesting that powerful environmental conditions (near death from starvation, for instance) can leave an imprint on the genetic material in eggs and sperm. **These genetic imprints can short-circuit evolution and pass along new traits in a single generation.**

A decade after the publication of Cairns's paper, professor of biology at Indiana University P. L. Foster wrote, "Much subsequent research has shown that mutation rates can vary, and that they increase during certain stresses such as nutritional deprivation. The phenomenon has come to be called "*adaptive mutation*." Today, adaptive mutation has been transformed into epigenetics. And suddenly, every university lab is pursuing it.

A favorite animal that geneticists love to study is C. elegans. Between October 1994 and January 1995, seventy-three scientific articles about C. elegans appeared in international journals. C. elegans is a very primitive worm about 1 mm in length that lives in the soil (FIG 1.5). C. elegans is an appealing and effective model organism for research because it is easy to work with in the lab, requires little food, and produces a large number of offspring by self-fertilization within a few days.

The worm is conceived as a single cell that undergoes a complex process of *morphogenesis*.[1] It has a nervous system with a "brain" (the circumpharyngeal nerve ring). It exhibits behavior and is even capable of rudimentary learning. C. elegans produces sperm and eggs, mates, and reproduces. All 959 somatic cells of its transparent body are visible with a microscope, and its average life span is a mere two to three weeks. Importantly, worms and humans share up to 80 percent of their genes. Not surprisingly, approximately half of all the known genes that are involved in human diseases can also be found in C. elegans. Scientists delight experimenting on this creature.

For example, researchers at Duke University have conducted a new study on the effects of starvation. What they did was to starve one group of C. elegans roundworms for one day and another group for eight days at the first stage of larval development after hatching. When feeding was resumed, the worms that were starved longer grew more slowly, and ended up smaller and less fertile. They also proved more susceptible to a second bout of starvation. Their offspring were smaller, fewer, and less fertile. However, these children and grandchildren of famine turned out to be more resistant to starvation, as if they had a memory of famine.

The field of epigenetics gained momentum when several decades ago scientists studied the children born to women who were pregnant during a period of famine toward the end of World War II in the Netherlands. They found that these children carried a particular chemical mark, or epigenetic signature, on one of their genes. The researchers linked that finding to differences in the children's health later in life. The children grew smaller than the Dutch average and had higher than average body mass. Their children were also smaller and more susceptible to diabetes, obesity, and cardiovascular disease. These changes were detectable over three subsequent generations.

It is not just food that can starve offspring. In humans, so can poverty—as demonstrated by a British study at the University of Bristol. The researchers selected forty men from a group of three thousand born in 1958—half born into rich households and half born into poor ones. In the study, subjects were chosen from the top and bottom 20 percent according to socioeconomic status, so ensuring they had examples of both extremes.

Focusing on stretches of DNA called *promoter regions*, which translates to *switches*, the team examined more than 20,000 sites throughout the genome. The patterns were different between the two groups on almost one third of the sites. Most tellingly, methylation levels were drastically different at 1,252 sites of the men who came from poor households, but only at 545 sites in men from rich families. Because the samples were taken in middle age, the researchers couldn't tell exactly when the epigenetic methyl groups were added or subtracted. While it is possible that the genes were altered in infancy, childhood, or even adulthood, the scientists conducting the experiments were of the opinion that the epigenetic changes they observed in adult DNA were largely the result of early life experience.

Today the most common surgical procedure in fertile women is delivery by elective cesarean section. Therefore, pregnant women or those planning on starting families should be aware of the research that has shed light on the fact that children born by cesarean section are at increased risk of developing asthma, type 1 diabetes, obesity, celiac disease, cancer, and suppression of their immune response.

Investigating this phenomenon, molecular cell biologists at the renowned Karolinska Institute in Sweden studied epigenetic alterations in the cord blood taken from elective C-section and vaginal delivery babies born at term. Blood stem cells from infants delivered by C-section were more

DNA-methylated than DNA from infants delivered vaginally. The researchers found specific epigenetic differences between the groups in 343 DNA regions, including genes known to be involved in processes controlling metabolism and immune deficiencies. These studies clearly indicate that the epigenome of an infant is sensitive to the prenatal environment and the experience of birth.

During the recent COVID-19 pandemic it was observed that some people became severely ill, while others tested positive with the virus but remained symptom free. The multigenerational weakening of the immune system described here may be a hitherto unrecognized factor, along with socioeconomic ones, which of course cause stress, physical and mental, accounting for these wide variations.

Complex diseases such as cancer, diabetes, obesity, autism, and birth defects are increasing in prevalence at rates that cannot be explained by classical genetics alone. Studies in humans and animals strongly suggest that epigenetic mechanisms may be responsible.

Social Epigenetics

Using mice as a model to investigate human breast cancer, researchers have demonstrated that a negative social environment (in this case, isolation) causes increased tumor growth. The findings also support previous epidemiologic studies suggesting that social isolation increases the mortality of patients suffering of chronic diseases, as well as clinical studies revealing that social support improves the outcomes of cancer patients. The presence of compassionate and caring people can alter the level of gene expression in a wide variety of tissues including the brain. Of course, our recent experiences with the COVID-19 pandemic taught us firsthand the veracity of the human need for interaction.

One of the most beautiful and imaginative experiments that I have encountered in my professional life is the "Kidnapping-and-Cross-Fostering Study" devised by Gene Robinson, director of the Institute for Genomic

Biology at the University of Illinois. What Robinson did was to pluck about 250 of the young bees from two African hives ("killer bees") and two European hives (a gentle bee strain), and paint marks on the bees' backs to identify their origins. (I don't think Robinson painted many bees. That's what postgrads are for.) Then he and his team switched each set of newborns and placed them into the hive of the other subspecies.[2] European honeybees raised among more aggressive African bees not only became as belligerent as their new hivemates—they came to genetically resemble them. And vice versa. What this experiment convincingly proves is that in a very short time the social environment can radically change gene expression and behavior.

David Clayton, a neurobiologist and a colleague of Gene Robinson at the University of Illinois, found that if a male zebra finch heard another male zebra finch sing nearby, a particular gene in the bird's forebrain would be stimulated and it would do so differently depending on whether the other finch was strange and threatening or familiar and safe. Songbirds demonstrated massive, widespread changes in gene expression in just fifteen minutes.

We are learning that brain responses to social stimuli can be massive, involving hundreds, sometimes thousands of genes. Even as recently as twenty years ago no self-respecting geneticist or neuroscientist would have thought in their wildest dreams that social experiences lead to changes in brain gene expression and behavior. Yet they do.

Trans-one-generation (F1)

The term *trans-one-generation epigenetics* refers to the effect maternal or paternal genetic factors exert on a child. Scientists designate these as F1 factors.

At one time or another, I am sure all of us have wondered why some people are always calm, no matter what, while others get anxious at the drop of a hat. Recent research on rats provides a clue. Mother rats seem to fall into two groups. Those that spend a lot of time licking, grooming, and nursing their pups, and others that simply ignore their pups. Highly nurtured rat pups tend to grow into calm adults, while rat pups deprived of nurturing care grow up anxious. The nurturing behavior of a mother rat during the first week of life shapes her pups' epigenome. And the epigenetic pattern that the mother establishes tends to endure, even after the pups become adults. **The difference between a calm and an anxious rat is not genetic; it is epigenetic.**

These data indicate that a higher level of maternal caretaking behavior during the first week of life promotes adult behavior that is characterized by stress resilience and increased maternal care in the offspring. In this instance, what is true for rats also applies to humans. Future parents, please take note.

In 2011 a University of Delaware group decided to study whether early life adversity alters gene expression. They exposed male and female infant rats to stressed "abusive" caretakers for thirty minutes daily during the first seven days of life. They induced abuse in mother rats by placing them in an unfamiliar environment with limited space. As a result, caretakers began to step on, drop, drag, actively reject, and roughly handle their infants. This treatment, naturally, elicited distress responses in the infants.

Brain-derived neurotrophic factor (*Bdnf*) regulates the survival and growth of neurons and influences synaptic efficiency and plasticity. The maltreated infant rats were found to have significant decrease in Bdnf gene expression that was in line with previous findings where early life experiences are known to have a lasting impact on this gene, leading to detrimental changes in personality and behavior. This underperforming Bdnf gene persisted through development and into adulthood.

Abusive and neglectful caregivers are known to leave children particularly susceptible to cognitive and emotional dysfunction. Indeed, childhood maltreatment is significantly associated with the later diagnosis of adolescent and adult major depression, schizophrenia, borderline personality disorder, and post-traumatic stress disorder.

Sensitive caregiving, on the other hand, is of immense value to a child. For example, neurobiologist Regina Sullivan at the NYU School of Medicine found that rat infants experiencing pain had several hundred genes more active than rat infants free of pain. With their mothers present, however, fewer than one hundred genes were similarly expressed (activated). Sullivan has successfully demonstrated that a mother comforting her infant in pain alters gene activity in a part of the brain involved in emotions (amygdala) and thus elicits a positive short-term behavioral response in her child.

This has important implications for understanding the biology of attachment and bonding in the postnatal period. Every expression of a mother's affection changes gene activity in her infant's brain, which leads the infant to gradually develop attachment to the mother. The smiling child's response elicits epigenetic changes in the mother. The repetition of this interaction over time eventually contributes to the development of mutual love: the love of the mother for her child (bonding) and the child's love for their mother (attachment).

Trans-two-generation (F2 and F3)

F2, F3 epigenetics is used here to refer to measuring the effects of parental behavior, stress, or trauma on their children, grandchildren, and great-grandchildren; i.e., first-, second-, and third-generation effects, F1, F2, and F3. These effects are often sex-specific.

I admire the following study from Emory University. It is beautifully simple and straightforward. The researchers tested a certain distinct smell experience of parents on the behavior and brains across generations of their progeny. The scientists gave male lab mice electric shocks every time they were exposed to the smell of acetophenone, a chemical used in perfumes. As a result of this classical conditioning technique the mice became anxious at the mere scent of acetophenone. **Their children came to fear the smell too, even though they had never been exposed to it.**

The mice showed no reaction to other smells and had no fear responses to sounds or different types of warnings. To confirm this, the scientists even took sperm from the first set of mice, then used in vitro fertilization (IVF) techniques to implant the sperm in females from another lab. The pregnant mice were raised in isolation, away from any contact with other mice, and yet their children still demonstrated an increased sensitivity to the original scent.

Similar results were achieved by Australian researchers studying mice infected with the toxoplasma parasite. They discovered that sperm of infected fathers carried an altered "epigenetic" signature that impacted the brains of resulting offspring. Molecules in the sperm called microRNAs (miRNAs) appeared to influence the offsprings' brain development and behavior (see p. 20 for more on miRNA). Transgenerational inheritance of similar epigenetic modifications has been associated with neuropsychiatric dysfunction in these mice's children and grandchildren.

Other researchers studied the effect of stress in four generations of rats. They found that a single exposure to prenatal stress in mothers increased the risk of preterm birth and adversely affected their offspring in many areas. The scientists involved in the study emphasized that **the causes of many complex diseases are likely rooted in the experiences of our ancestors.**

Exposure of mothers to nicotine and other components of cigarette smoke is recognized as a significant risk factor for behavioral disorders, including attention deficit hyperactivity disorder (or ADHD) in many generations of descendants. To study whether the same applies to fathers, researchers at Florida State University in Tallahassee exposed male mice to low-dose nicotine in their drinking water during the stage of life in which the mice produce sperm.

They then bred these mice with females that had never been exposed to nicotine. While the fathers were behaviorally normal, both sexes of offspring displayed hyperactivity, attention deficit, and cognitive inflexibility. When female (but not male) mice from this generation were bred with males never exposed to nicotine, male offspring displayed fewer, but still significant, deficits in cognitive flexibility.[3] Analysis of spermatozoa from the original nicotine-exposed males indicated that multiple genes had been epigenetically modified, including the dopamine D2 gene, critical for brain development and learning, suggesting that these modifications likely contributed to the cognitive deficits in the descendants. These findings underscore the need for more research on the effects of smoking by the father, rather than just the mother, on the health of their children.

How about grandparents who smoke? Well, we have a study on that, too. After analyzing data from more than 14,500 children born in the United Kingdom during the 1990s, epidemiologists from the University of Bristol found that people with a maternal grandmother who smoked during her pregnancy had a 53 percent increased risk of developing autism. The study also revealed that girls whose maternal grandmother smoked during pregnancy were 67 percent more likely to have autism-linked traits. For two of the traits (social communication and repetitive behavior) the investigators demonstrated that granddaughters were much more affected than grandsons.

Today, in most parts of the world marijuana is perceived as benign and less harmful than alcohol. Duke University researchers have proven this belief unfounded. They analyzed differences between the sperm of males who smoked or ingested marijuana compared to a control group with no such experiences. They identified significant hypomethylation in the sperm of men who used marijuana compared to controls, at a gene that has been strongly implicated in autism, schizophrenia, and post-traumatic stress disorder. This hypomethylated state was also detected in the forebrain region of rats born to fathers exposed to THC, giving rise to cognitive deficits.

Researchers from the Mount Sinai School of Medicine in New York decided to study more psychological issues with far-reaching results. They subjected adult male mice to chronic social defeat stress. Then they bred the stressed mice and a control group of male mice with normal female mice. Once their offspring were born, they were assessed by a variety of standard tests for depressive and anxiety-like symptoms. Plasma levels of corticosterone and vascular endothelial growth factor were also measured in the two groups of infant mice.

Both male and female offspring from the defeated fathers demonstrated pronounced depressive and anxiety-like behaviors. The offspring of defeated fathers also displayed increased basal levels of plasma corticosterone and decreased levels of vascular endothelial growth factor, both of which have been implicated in depression. These and related findings in mice show that part of an individual's risk for clinical depression or other stress-related disorders may be determined by his or her father's life exposure to stress.

Studies curated from many leading research institutions confirm that intense experiences encountered in a parent's lifetime may, utilizing mechanism of transgenerational inheritance, impact the physical and mental health of their children for countless generations. Though hard to prove at the moment, a rising tide of supporting research seems to be carrying forward the corollary hypothesis; namely, that transgenerational inheritance also is responsible for many cases of unexplained phobias, anxieties, mood disorders, or personality traits.

Trauma

Let us now move to a subject that is much in the news today; namely, trauma. Trauma is generally defined as an event that induces severe fear, helplessness, or horror. *Post-traumatic Stress Disorder (PTSD)* is a particular type of trauma. Essentially, PTSD occurs when a person feels overwhelmed and helpless in a life-threatening situation. As witnessed by many veterans, PTSD has long-lasting and often debilitating effects.

In the last fifty years, the transgenerational transmission of trauma has been explored in more than five hundred articles. Many studies question the concept, others support it. Child psychoanalyst Anna Freud first described transgenerational trauma in 1942. In the same year, Dorothy Burlingham, an American child psychoanalyst, educator, and lifelong friend and partner of Anna Freud, referred to unconscious "messages" passed between mothers and children during the German bombing of London in World War II.

Margaret Mahler, a child psychoanalyst in the United States, observed in 1968 that in early infancy a mother and her child function almost as one psychological unit. She held that there is fluidity between a mother's and a child's psychic borders. Two decades later, a Lakota professor of social work named Maria Yellow Horse Brave Heart coined the phrase *historical trauma*. What she meant by that was "the cumulative emotional and psychological wounding over the life span and across generations."

As demonstrated by the above brief historical retrospective, in the past, psychologists and psychiatrists thought of transgenerational transmission of trauma in terms of a purely psychological phenomenon. According to this theory, starting at conception the mother's anxiety, unconscious fantasies, perceptions, and expectations are passed to the child's mind and body by verbal and nonverbal cues. Parents who have experienced trauma may constantly talk about it or, as is more often the case, never talk about it. Just living with a person (survivor or veteran) who suffers from PTSD can be traumatizing. The children in such families experience their own PTSD by "walking on eggshells" around the PTSD parent and wondering what they are hiding.

The idea that a parental traumatic experience could be passed on to subsequent generations gained acceptance in scientific circles in the late '70s and early '80s. Since the mid-1980s controlled studies on the children of Holocaust survivors (in reality, adults) showed increased vulnerability to PTSD, distrust of the world, impaired parental function, chronic sorrow, inability to communicate feelings, an ever-present fear of danger, separation anxiety, boundary issues, and other psychiatric disorders.

Today, we are learning that the spoken or unspoken messages of PTSD parents may impact the child both on a biological as well as a psychological level. In the same way parents pass on genetic characteristics to their children, they also pass on all kinds of "acquired"; that is, epigenetic characteristics, especially if these originated in powerful emotionally charged experiences such as exposure to starvation, violence, or the tragic loss of loved ones. Such traumatic events leave an imprint on the genetic material in germ cells of individuals and may be transferred to their children and their children's children.

Rachel Yehuda, professor of psychiatry and neuroscience at the Mount Sinai School of Medicine, has undertaken to study the children of Holocaust survivors. She found that they were three times more likely to develop post-traumatic stress disorder when they were exposed to a traumatic event than demographically comparable parent and offspring control subjects. Furthermore, these children exhibited the same neuroendocrine (hormonal) abnormalities that were observed in Holocaust survivors and persons with post-traumatic stress disorder.

Following the September 11, 2001, terrorist attacks, Yehuda and her colleagues performed a longitudinal study on thirty-eight women who were pregnant on 9/11 and were either at or near the World Trade Center at the time of the attack. The children of women who were traumatized as a result of 9/11 subsequently manifested an increased distress response when shown novel stimuli. Children with the largest distress response were the ones born to mothers who were in their second or third trimester when exposed to the World Trade Center attacks.

Neuroscientists at the University of Zurich explored the effect of early trauma by artificially separating male mice from their mothers at unpredictable times in the first two weeks of life. When these young mice became adults, they were more hesitant to enter open spaces and brightly lit areas than mice that had not been separated from their mothers. If they were people, we would diagnose them as neurotic. These behavioral changes were present in the mice's offspring, which also displayed alterations in metabolism, and in their offspring's offspring. **This study successfully demonstrated for the first time that traumatic experiences affect metabolism and that these changes are hereditary.**

Multiple rodent and nonhuman primate studies have also shown that **early trauma produces lasting changes in neural function and behavior.** One mediator of this process may be the before mentioned *Bdn protein*, which produces changes in the *Bdnf gene.*

How Epigenetic Changes Are Passed On

Most of the research in the field of epigenetics has focused on epigenetic mechanisms involving DNA and certain molecules (methyl groups and acetyl groups) that attach to DNA. There is much discussion among geneticists as to how epigenetic changes are passed on through sperm and ova. Now scientists are discovering that the classical genetic code is not the only code involved in the regulation of cell differentiation and behavior in multicellular organisms. There is a second level of control that contributes to the regulation of gene activity. One of these is based on chemical modifications of the histone proteins. I briefly mentioned histones at the beginning of this chapter. Histones have attracted relatively little attention until now. Histones are distinct from DNA, although they combine with it during cell formation, acting a bit like a spool around which the DNA winds (FIG 1.2).

A new collaborative study by McGill University and Swiss researchers have discovered that histones are part of the content of sperm transmitted at fertilization. The researchers created mice in which they slightly altered the biochemical information on the histones during sperm cell formation. The offspring for two successive generations were adversely affected both in terms of their development and in terms of their survival.

These findings are remarkable because they indicate that information in addition to DNA is involved in heritability. **The study highlights the critical role that fathers play in the health of their children and even grandchildren.**

Another proposed mechanism for gene regulation involves small non-coding RNAs called *micro RNAs (miRNAs)* found in many mammalian cell types including sperm. About 60 percent of the genes of humans and other mammals appear to be targeted by miRNAs. miRNAs constitute a newly discovered type of gene regulator, where each miRNA controls a distinct set of genes. miRNAs are proving to be master regulators of virtually all cell processes with broad controls stretching into cell cycle, signal transduction, and energy metabolism pathways, among others.

In a Tufts University School of Medicine study, male mice exposed to chronic social instability stress during adolescence transmitted stress-associated behaviors to their female offspring across at least three generations even if they never experienced significant stress themselves or interacted with their fathers by way of the male lineage. One mechanism for this effect was found to be sperm miRNA.

Sperm miRNA expression in humans has been known to be affected by environmental factors, such as smoking and obesity. However, a University of Delaware group was first to demonstrate sperm miRNA changes in response to stress in humans, and raised the possibility that sperm miRNA could be a biomarker for early abuse as well as elevated susceptibility of offspring to psychiatric disorders.

miRNAs play an important role in defending the body from invasions by viruses, as we were so tragically reminded when the COVID-19 crisis erupted. They do so by latching onto and cutting the RNA material of the virus. One of the reasons that the COVID-19 virus has had such a devastating effect on older people and those with underlying conditions is because as we age and develop chronic medical diseases, our miRNA numbers dwindle, reducing our ability to destroy invading viruses.

Further new details about miRNAs have been successfully unraveled by neurobiologists at the University of Maryland studying extracellular vesicles. The male reproductive tract, the *caput epididymis*, the structure where sperm matures, is where these tiny vesicles packed with miRNAs originate. Vesicles fuse with sperm to change its payload delivered to the egg. The caput epididymis responds to the father's stress by altering the content of these vesicles.

Extracellular vesicles have emerged as important mediators of intercellular communication, involved in the transmission of biological signals between cells. They have been identified as regulating a diverse range of biological processes, effects of stress as well as contributing to the development of infectious diseases, cancer, and neurodegenerative disorders. Extracellular vesicles help pass information between cells and onto offspring. In an important departure from past assumptions, the scientists leading the Maryland study **have now accepted the idea that sperm can be vulnerable to environmental factors.**

Similarly, evidence is accumulating that preconceptional exposure to certain lifestyle factors, such as diet, physical activity, and smoking affects the development of the next generation through alterations of the epigenome of sperm cells.

A related study at the University of Massachusetts Medical School showed that embryos fertilized using sperm from the distant portion of the epididymis—where sperm have not yet gained a full payload of regulatory RNAs—exhibit gene dysregulation early in development and then fail to implant in the uterus efficiently. Clearly, small RNAs in sperm are essential for a healthy pregnancy.

In the last decade, scientists have established that small RNA molecules can be found outside cells in blood, urine, tears, cerebrospinal fluid, breast milk, amniotic fluid, seminal fluid, and others. Moreover, scientists have discovered that small bits of circulating RNA can reflect particular conditions, such as the presence of a cancerous tumor or pregnancy-related disorders.

While some scientists remain skeptical that extracellular RNA and DNA are anything more than debris, a combined team of neurogeneticists from the University of Oxford and Massachusetts General Hospital regard them as a **newly discovered form of communication among cells** that plays a significant role in human health. For example, mutiple studies suggest that small RNAs act as instructions that help coordinate an immune response.

I am in total agreement with Marcus Pembrey, emeritus professor of pediatric genetics at University College London, who has been championing the idea of epigenetic inheritance for over a decade. Pembrey has said, "It is high time public health researchers took human transgenerational responses seriously. I suspect we will not understand the rise in neuropsychiatric disorders or obesity, diabetes, and metabolic disruptions generally without taking a multigenerational approach."

Personal Epigenetics

For a long time, we have known that the mind affects the body and conversely, that the body affects the mind. For example, people under stress are more likely to become ill and people who suffer from the flu or other health issues often feel blue or even depressed. On the other hand, pursuing activities that engage you fully, searching for meaning outside yourself, having friends, being married, and being intimate are all strongly associated with happiness, which in turn is associated with good health. In medical parlance this reciprocal process has been referred to as psychosomatic medicine or body-mind medicine.

Another way in which the mind can affect the body is through the widespread practice of meditation. According to a study undertaken in Wisconsin, meditating for eight hours on mindfulness, altruistic love, and compassion induces major epigenetic modifications. Compared to a control group whose members did not meditate but engaged in leisure activities in the same environment, the researchers found that the meditators were more resilient to infections and diseases in general than the control group. The meditators achieved positive, highly beneficial **epigenetic transformation by way of self-regulation.** We would all benefit, I think, if we adopted the results of this study to our own lives.

Let us look at another study on self-regulation. Researchers divided eighty-four hotel maids in New York into two groups. One group was told that the work they do on their job is good exercise and satisfies the Surgeon General's recommendations for an active lifestyle. This information was the equivalent of a placebo. The other group (control) was not given this information. Although actual behavior did not change, four weeks after the start of the experiment, the informed group perceived themselves as getting significantly more exercise than before. Compared to the control group, they showed a decrease in weight, blood pressure, body fat, waist-to-hip ratio, and body mass index. The women in the experimental group did not work harder than the controls but **their belief system actually changed the way their bodies functioned**.

Examining these studies in aggregate, they clearly illustrate that our thoughts and feelings are much more powerful than we realize. A thought is a set of neurons firing that, through complex brain wiring, will activate multiple intersecting pathways, emotional and pain centers, memories, the autonomic nervous system, the genome, and other parts of the embodied mind. The genome responds equally to both: external; that is, environmental stimuli and internal stimuli (thoughts, feelings, moods) with epigenetic modifications.

The Good News

According to a new study from the University of Helsinki, classical music fans, when listening to Mozart's Violin Concerto No. 3 in G major, were found to upregulate the activity of their genes involved in dopamine secretion and transport, synaptic neurotransmission, learning and memory, and downregulate the genes mediating the destruction of neurons, which is all for the good. What this means is that if you find something pleasurable, it can change your gene expression. Not Mozart, per se.

How we live our lives can have significant effects on how we age and develop diseases, including cancer. On the physical side of the equation, if we look at colon cancer, researchers from the University of Basel found that aspirin and hormonal replacement therapy reduced the methylation rate of colon cancer–related genes, whereas smoking and high body mass index (BMI) increased it.

Steve Cole, professor of medicine, psychiatry, and biobehavioral sciences at the UCLA School of Medicine, has written much on the subject of self-regulation. He holds, and I totally agree with him, that we are architects of our own lives more than we realize. Our subjective experience carries more power than our objective situation. If we feel good about ourselves, not only will our health improve but so will our relationships. Others will like and respect us, which in turn will make us feel even better about ourselves. Thus, we create a self-reinforcing reward system grounded in epigenetics.

Being optimistic also helps. A thorough review of the medical literature to determine the strength of the association between optimism and physical health revealed that optimism was a significant predictor of health outcomes in cardiovascular disease, including immune function, cancer, complications related to pregnancy, and physical symptoms such as pain. People who feel enthusiastic, hopeful, and cheerful—what psychologists call "positive affect"—are less likely to experience memory decline as they age. It does not necessarily mean they will never get ill (mentally or physically), but optimists diagnosed with bipolar illness are able to manage the disease better than pessimists. The same applies to people suffering from depression. All these and many more studies add to a growing body of research on the contribution an optimistic outlook makes to health.

Of course, I am not suggesting a "fake it till you make it" attitude. Cultivating creativity, imagination, self-reflection, and living a meaningful and engaged life takes work but is an investment in our overall well-being—and potentially the well-being of our children.

Summary

In the 1850s, when Darwin first advanced his theory of natural selection and survival of the fittest, the underlying molecular mechanisms of genetics were unknown. However, over the past fifty years, advances in genetics and molecular biology have led to a neo-Darwinian theory of evolution based on epigenetics. Our survey of recent discoveries in epigenetics has made it abundantly clear how nature (genes) and nurture (the environment) work in concert. It is not one or the other that is responsible for a disease or personality trait. The only thing we know for sure is that we are the product of a dynamic interaction between these forces and that nothing about us is written in stone. Therefore, as long as we breathe we are a work in progress, constantly changing. Epigenetic modifications are dynamic and potentially reversible processes that take place over our entire lifetime.

In view of the above cited research—which does not claim to be exhaustive but rather representative of the field—there is robust biological evidence of transgenerational transmission of trauma to offspring by both, fathers and mothers. Probable familial factors are miRNAs and lncRNAs (long noncoding RNAs), as well as epigenetic changes in maternal and paternal germ cells. In the case of fathers, extracellular vesicles play an important role. In mothers there is the insulin-like protein and programming of the HPA axis of her baby during pregnancy.[4]

An epigenetic hypothesis for environmental contributions to physical and emotional health continues to gain traction. Even more significantly, clear and convincing evidence from basic and clinical research at leading universities indicates that an organism adapts to changes in its environment through alterations in their gene expression. Consequently, before their offspring is even conceived, parental life experiences and environmental exposures modify their germ cells and in turn affect the development and health not only of their children, but that of their grandchildren and great-grandchildren. Similarly, children's fears, and their children's fears, anxieties, and personality attributes, may be affected by their parents' mindsets.

In addition, compelling scientific data show that our social lives, interactions of others and ourselves, can change our gene expression with a rapidity, breadth, and depth previously unknown. Genes don't make us who we are. Gene expression does. And gene expression varies depending on the life we live. In other words, the food we eat, the water we drink, the air we breathe, our interpersonal relationships, and our relationship to ourselves—they all affect us on a deep biological level, which in turn affects our minds.

Do genes matter? Absolutely. But so does the physical, psychological, and social environment, not only from birth on but extending back to the nine months of womb life, conception and, in many ways, several generations further back. The advance of epigenetics has shattered the old Darwinian paradigm of genetics.

KEY TAKEAWAYS

- An individual's adult physical and mental health is heavily influenced by their early prenatal environment.
- The unborn child will adjust as best they can to the external environment they are going to encounter upon birth by way of prenatal epigenetic changes.
- Generally speaking, the life experiences of parents may impact the development and health of their descendants.
- In particular, parental nicotine, cannabis, and alcohol use has been shown to be associated with adverse neurodevelopmental outcomes in offspring.
- Traumatic experiences of parents lead to extra-sensitivity to traumatic events in offspring, and this may persist for several generations.
- Gene activity increases or decreases in response to changes in our environment.
- Our interactions with others and ourselves rapidly lead to changes in brain gene expression and behavior.
- Genes don't make us who we are. Gene expression does. And gene expression varies, depending on the life we live.

CHAPTER TWO

THE BRAIN: HOW IT REMEMBERS WHAT IT REMEMBERS

Introduction

Learning and memory are two of the most remarkable faculties of our mind. Learning is the biological process of acquiring new knowledge about the world, and memory is the process of retaining, reconstructing, and accessing that knowledge over time.

One of the most challenging problems in neuroscience is: How do short-term chemical changes lead to something long term, like memory?

Ask most people where in their body memory resides and they're most likely to look at you as if to say, "What a dumb question! In the brain, of course." Well, maybe not. In fact, there are instances where the brain cannot possibly account for what we know and how we function in the world.

In this and the following chapters I shall offer extensive research that disputes the widely held cortico-centric hypothesis of memory and the mind. Instead, I shall advance a more holistic body-mind explanation, the embodied mind theory.

Neuroscience in a Nutshell

The average adult brain weighs about three pounds. It is made up of around 75 percent water. The brain consists of roughly 100 billion neurons, as many as the stars in our galaxy, embedded in a scaffolding of a 100 billion glial cells. Each neuron may have 1,000–10,000 synapses (connections with other neurons). The most active period of neuron proliferation takes place during the middle of the second trimester, when 250,000 neurons are created every minute.

There are no pain receptors in the brain, so the brain does not feel pain. There are 100,000 miles of blood vessels. While in the past it was held that we are born with all the neurons we will ever have, now we know that new neurons are created every day.

Early experiences have **decisive impact** on the architecture of the brain, and on the nature and extent of adult capacities. Brain development is **non-linear**: there are prime times for acquiring different kinds of knowledge and skills. By the time children reach age three, their brains are **twice as active** as that of their pediatrician. However, if you have young children, be warned: their brain activity levels will drop during adolescence.

The nervous system in our bodies consists of the central nervous system (CNS) and the peripheral nervous system (PNS) that in turn gives rise to the autonomic nervous system (ANS) which is divided into the sympathetic and parasympathetic systems. The autonomic nervous system is a control system that acts largely unconsciously and regulates bodily functions such as the heart rate, digestion, respiratory rate, pupillary response, urination, and sexual arousal.

The sympathetic nervous system is the "fight-or-flight" system that controls our response to stress and originates in the thoracic spine. (The *thoracic spine* connects the *cervical spine* above with the *lumbar spine* below.) The parasympathetic system is the "rest and restore" or "feed and breed" system. It regulates smooth muscle contraction and originates in the head and the *sacral* region of the spine (a triangular bone in the lower back formed from fused vertebrae and situated between the two hipbones of the pelvis).

The basic unit of the central nervous system is the *neuron* or nerve cell. Each neuron (FIG 2.1) has several thousand *dendrites*—up to ten thousand—tiny hairlike strands of tissue that receive signals, and one *axon*, a more robust structure through which the neuron sends signals to other cells. Neurons do not actually touch. Each axon produces about 160 different neurotransmitters that cross a minuscule gap, the *synapse* (FIG 2.3) to insert themselves in the receptors of dendrites that are structured to receive a particular neurotransmitter and no other. Sort of like a space shuttle docking to a space station.

I am sure you have heard of Prozac, the first antidepressant of a new class of drugs called serum serotonin reuptake inhibitors (SSRIs). What Prozac and the other SSRIs do is that they occupy some serotonin receptor sites on dendrites. When the axon produces serotonin, it has no place to go because another space capsule, Prozac, has occupied its docking station. Consequently, serotonin concentration increases in the brain. Many scientists believe that depression is due to low serotonin levels—though no one knows the cause of that. I am not a great fan of psychotropic drugs (drugs for depression, anxiety, and other emotional problems), but I will say that I have seen stunning improvements in some patients as a result of taking these drugs.

Neurons differ from other cells in the body by their ability to carry electric signals and to transfer chemical signals to other neurons. Neurons function in networks. And this incredibly complicated network, composed of billions of connections, is called the *connectome* (FIG 2.2).

The human brain's advanced cognitive capabilities are attributed to our recently (the past one hundred thousand years) evolved *neocortex*. Comparison of human and rodent brains shows that the human cortex is thicker, contains more white matter, has larger neurons, and its abundant pyramidal cells (the cells that do most of our thinking) have more synaptic connections per cell as compared to rodents. A team of researchers led by Professor Idan Segev from the Hebrew University of Jerusalem, took direct measurements of membrane capacitance in human pyramidal neurons. Segev demonstrated in his work that human cortical neurons are efficient **electrical microchips**, compensating for the larger brain and larger cells in humans, and processing sensory information more effectively. The concept of cells as microchips is central to the Embodied Mind Hypothesis and will be further explored in the next chapter.

There is good evidence that specific neurotransmitters such as epinephrine, dopamine, serotonin, glutamate, and acetylcholine are involved in the formation of memory. Although we don't yet know which role each neurotransmitter plays in memory, we do know that communication among neurons by way of neurotransmitters is critical for developing new memories.

It is also believed that strong emotions trigger the formation of lasting memories, and weaker emotional experiences form weaker memories. This is called arousal theory.

Neurons comprise only 15 percent of the brain. The other 85 percent is made up of *glial cells*. Glial cells continue to grow in number until a few years after birth. They guide early brain development and keep the neurons healthy throughout life. Glial cells provide the scaffolding for neurons and as the origin of their name implies (Greek for "glue"), they help to keep the neurons bonded. Glial cells can affect the functioning of neurons even though they cannot discharge electrical impulses of their own.

Human neurons are very similar to those of other animals, right down to using the same neurotransmitters. But as one compares the brains of animals ascending the evolutionary tree, one sees that the higher you climb, the more nonneuronal glial cells these animals' brains contain in proportion to the number of neurons. For years glial cells were dismissed as mere putty, just as noncoding DNA was considered junk. Actually, glial cells control communication between neurons and play an essential role in learning. Glial cells form three subgroups; *oligodendrocytes, microglia, and astrocytes.*

Where Do We Keep Our Memories?

On the following pages I shall first describe the accepted scientific hypothesis regarding memory and then cite research that shatters the old paradigm I call "The Bedrock Theory of Learning and Memory."

According to this theory, incoming signals from our sense organs initiate the production of specific proteins in neurons that make the synapses (FIG 2.3) grow stronger. These proteins not only mold and shape the synapse but also encode memories. Just like physical exercise leads to greater muscle mass through the production of new proteins, so experience builds memories in synapses, potentially whole neural networks and brain regions. The general idea that learning induces modification of synapses that result in the storage of memories in an ever-changing plastic brain has become one of the dogmas of modern neuroscience and is usually presented in the scientific literature as well as in the popular press as an established fact and "generally accepted."

According to this reigning neuroscientific view, short-term memory is linked to functional changes in existing synapses, while long-term memory is associated with a change in the number of synaptic connections and strengthening of the brain's existing circuitry. As we shall see, this is a highly problematic notion.

The frontal cortex (FIG 2.4) can tap into the sensory information immediately for use as a short-term, or working, memory. The hippocampus and areas of the medial temporal lobes begin to encode this new information into a long-term memory by growing new neural connections and strengthening the brain's existing circuitry. Retrieval of emotionally charged memories occurs by way of the amygdala, the hippocampus, and prefrontal cortex system.

Short-term and long-term memories reside in different parts of the brain. If you were to stimulate one area of the brain such as the occipital cortex at the back of the brain with a tiny electrical probe you would trigger visual memories; and the left temporal area, at the side of the brain, might produce speech sounds, words, phrases, etc. Related memories are stored in adjoining regions. Stimulation of larger fields will lead to the emergence of more complete memories. For the recall of experiences and facts, various parts of the brain have to work together. Much of this interdependence is still undetermined. However, it is widely held that memories are stored primarily in the cerebral cortex and that the control center that generates memory content and also retrieves it is located in the brain's middle axis.

Eric Kandel is a professor of biochemistry and biophysics at Columbia University who shared the Nobel Prize in the year 2000 with Arvid Carlsson and Paul Greengard for "their discoveries concerning signal transduction in the nervous system." Kandel conducted his studies on the marine snail *Aplysia* (FIG 2.5), which has only about twenty thousand nerve cells compared with about a one hundred billion in the human brain. The snail has a simple reflex by which it protects its gills and Kandel used that reflex to study how the snail learned and remembered stimuli. He showed that short-term memory involves increased levels of neurotransmitters at the synapses (the communication sites between nerve cells), and long-term memory requires changes in the levels of proteins in the synapse (FIG 2.3). After learning how these simple animals functioned, he then experimented on mice. This work helped him understand how the same processes that occurred in nerve cells of slugs could be seen in mammals, which includes humans.

Kandel concluded that the basic building block of memory is the synapse, where both pre- and postsynaptic elements, together with associated glial processes, form an integral unit with an individual identity and distinct "neighborhood." The increase in connectivity strength within a diffuse group of cells in a more complex feed forward circuit results in the emergence of an engram (engrams are complex stored memories, (FIG 2.6) within a cell assembly.

I applaud Eric Kandel, whom I respect and heartily agree with, on his assertion in 2006 that "in the study of memory storage, we are now at the foothills of a great mountain range . . . To cross the threshold from where we are to where we want to be, major conceptual shifts must take place." However, I take exception to his theory that synapses store memories. In 2007 Stefano Fusi and Larry Abbott called for **"radical modification of the standard model of memory storage,"** and in 2012 Stuart Firestein echoed this in his book *Ignorance: How It Drives Science*, which also called for a departure from Kandel's hypothesis.

Recent Discoveries

Interestingly, it is recent work in this exact domain that has put into doubt the idea of synaptic conductance as the basic memory mechanism. David Glanzman's group at the University of California, Los Angeles, exposed Aplysia to mild electric shocks, creating a memory of the event expressed as new synapses in the brain. Then they transferred neurons from the mollusk into a petri dish and chemically triggered the memory of the shocks in them. Next, they added propranolol to the neurons. The drug wiped out the mollusk's synapses formed during learning. When the neuroscientists examined the brain cells, they found that even when the synapses were erased, molecular and chemical changes indicated that the engram, or memory trace, was preserved. These studies suggest that **memories are stored inside of neurons** in Aplysia and, very likely, in all animals.

In the same vein, researchers at the University of Pennsylvania have discovered in the mouse brain that a key metabolic enzyme, called *acetyl-CoA synthetase 2*, or ACSS2, works directly within the nucleus of neurons to turn genes on or off when new memories are being established.

Using mouse models, researchers in the laboratory of Carlos Lois at Caltech determined that strong, stable memories are encoded not, as has been up to recently postulated, by strengthening of the connections to an individual neuron but rather by teams of neurons all firing in synchrony.

Finally, Patrick Trettenbrein from the Language Development and Cognitive Science Unit, University of Graz, Austria, in a 2016 paper titled "The Demise of the Synapse as the Locus of Memory: A Looming Paradigm Shift?" reviewed the evidence and concluded that the **synapse is an ill fit when looking for the brain's basic memory mechanism.** It has been repeatedly shown that memory persists despite destruction of synapses and synapses are turning over at very high rates even when nothing is being learned. Taking into consideration all of the preceding, the case against synaptic plasticity is convincing.

To enable cognition and the storage of memories, the present, most evidence-based scientific view is that interactions between three moving parts—a binding protein, a structural protein, and calcium—are necessary for electrical signals to enter neural cells and remodel the *cytoskeleton*. Cytoskeleton is a dense network of various filamentous proteins in all cells with a nucleus, like humans and other animals have, which are essential for cell shape, cell division, and cell migration. *Actin filaments, microtubules,* and *intermediate filaments* form the major components of the cytoskeleton. It is in these cytoskeletons inside neurons where some of the leading scientists in the world believe that memories are stored. All cells in the body contain cytoskeletons, as we shall see in the chapter on the cell.

Several years ago, biologists realized that a single neuron could function as a logic gate, akin to those in digital circuits. Recently, researchers in Germany discovered that individual dendrites may process the signals they receive from adjacent neurons before passing them along as inputs to the cell's overall response. It seems that tiny compartments in the dendritic arms of cortical neurons can each perform complicated operations in mathematical logic. In theory, almost any imaginable computation might be performed by one neuron with enough dendrites, each capable of performing its own nonlinear operation.

A neuron behaving like a multilayered network has much more processing power and can therefore learn or store more data. "Very few people have taken seriously the notion that a single neuron could be a complex computational device," said Gary Marcus, a cognitive scientist at New York University. This discovery may also prompt some computer scientists to reappraise strategies for artificial neural networks, which have traditionally been built based on a view of neurons as simple, unintelligent switches.

Further research into how neurons perform their tasks by neuroscientist Jeffrey Macklis of Harvard Medical School, has challenged the dogma that the nucleus and cell body are the control centers of the neuron. What Macklis's results suggest is that growth cones—the outermost tips of the axonal strands—are capable of receiving information from the environment, making signaling decisions locally, and functioning semiautonomously without the cell body. The growth cones contain much of the molecular machinery of an independent cell, including proteins involved in growth, metabolism, signaling, and more. In a sense, our bodies are not top-down, hierarchically governed states but rather bottom-up, democratic societies where every cell and every part of a cell contributes to the greater good.

According to Douglas R. Fields, writing in *Scientific American*, "Neurons are elegant cells, the brain's information specialists. But the workhorses? Those are the glia." Interestingly, Dr. Fields's daughter, Kelly, produced the immunocytochemical staining and microphotography as it appears in figure 2.7. when she was in grade school and Dr. Fields took her to his lab to visit on "bring your daughter to work day." (At the present she is twenty-six years old and is a full-time rock-climbing guide, discovered through personal communication.)

Astrocytes (FIG 2.7), which are the largest glial cells, were long considered of little significance, sideline players in the brain. Not anymore. Although astrocytes cannot generate electrical impulses, they do communicate with one another and with neurons by way of the rise and fall of calcium concentrations. They also release *gliotransmitters*, substances chemically similar to neurotransmitters. Astrocytes not only sense communication between neurons at synapses, they can control neuronal communication.

In 2014 Professor Terrence Sejnowski, head of the Salk Institute's Computational Neurobiology Laboratory, and his colleagues, showed that disabling the release of gliotransmitters in astrocytes lowered a type of electrical rhythm known as a gamma oscillation, important for cognitive skills. In that study, when the researchers tested the learning and memory skills of mice with disabled astrocytes, they found them deficient in their capacity to discriminate novelty.

Human astrocytes are twenty times larger in volume than rodent astrocytes. This is far greater than the proportionate increase in size of human neurons relative to rodent neurons. The increase in number and complexity of astrocytes in the human brain contributes more than neurons do to the large increase in cerebral volume in humans and primates. The fact that human astrocytes are larger and more complex than those of other animals suggests that their role in neural processing has expanded with evolution.

Human astrocytes, rather than communicating like neurons by way of electrical signals, communicate with other astrocytes and with neurons using neurotransmitters. This allows them to pass signals three times faster than neurons.

A team of neuroscientists has grafted human brain cells into the brains of mice and found that the rodents' rate of learning and memory far surpassed that of ordinary mice. Remarkably, the cells transplanted were not neurons, but glia cells that are incapable of electrical signaling. The new findings suggest that information processing in the brain extends beyond the mechanism of electrical signaling between neurons.

Because of their size astrocytes span large numbers of neurons and millions of synapses and seem to contribute another level of functioning to neural networks. Therefore, it is not surprising that Argentinian neuroanatomist Jorge Colombo has spoken of **astroglia nets as a potential nonneuronal dimension of information processing, in which glia couple neurons and synapses into functional ensembles.**

An international study led by researchers from the National University of Ireland, Galway, demonstrated that the relationship between brain structure and intelligence not only involves gray matter but also the brain's white matter; in other words, the wiring system as a whole. As professor of psychiatry at Stanford University Robert Malenka so aptly put it, "When considering how the brain works, we need to analyze and understand all the different types of cells in the brain and how they interact."

The spinal cord, which like the brain is part of the CNS, consists of neurons and supporting glial cells. Researchers at the University of Montréal, using a new type of MRI machine, were, for the first time, able to show that the spinal cord retains learned motor movements independent of the brain. Not only that but the scientists showed that **the spinal cord might learn motor skills independently of the brain**. Of course, this study also demonstrates, rather persuasively, that neurons outside of the brain retain memories.

In 1998 the "Father of Gastroneurology," Michael D. Gershon, who teaches at Columbia University, jump-started research interest in the nervous system of the gut, the *enteric nervous system* (ENS), by writing the book *The Second Brain*. Since then, other scientists have found support for Gershon's hypotheses. Most recently, biologists at the Human Biology Department of the Technical University of Munich reviewed the evidence for the "smartness" of the ENS. They provided examples for habituation, sensitization, conditioned learning, and long-term facilitation. In their paper they put it this way, "Despite some remaining experimental challenges, we are convinced that the gut is able to learn, and are tempted to answer the question [Is the gut smart?] with: Yes, the gut is smart."

Breaking with Kandel's model, I conclude that the widely held belief that memory is stored in synapses is badly in need of an upgrade. The findings of Glanzman, Lois, Trettenbrein, and the other neuroscientists I have referred to indicate that synapses provide "access points" to neurons. Information flows through the synapses to the neurons. The more information of a particular type enters an assembly of neurons, the stronger become its synapses, and the more memory will be stored in the neurons and glial cells.

And now, I propose a little mind experiment. Imagine for a moment two villages separated by a dense forest. A few people walk through the forest between the villages and as they do they create a narrow path. Over time the population grows and more people walk along this path. The path gets wider. Then men on horseback, eventually even carriages begin to travel this way. The narrow path becomes a wide road now. The people who had traveled along the road will remember their experiences and may share these with others. On closer inspection, the road may reveal footprints or wagon marks but not any memory of the people and animals that traversed it. It is just a road.

I offer this little story as an analogy for the relationship of synapses to neurons. No doubt the synapses will get thicker as more information passes through them but the information will end up in the neurons and glial cells, and, as we shall see, in the rest of the body and not the synapses.

It is now well established that in addition to cortical neurons, glial cells, the neurons in the spinal cord and the gut, actively participate in regulating our bodies and minds.

When Neurons Go Missing

In the animal kingdom, vast ranges in brain size fail to correlate with apparent cognitive power. Crows and ravens, for example, have brains less than 1 percent the size of a human brain, but still perform feats of cognition comparable to chimpanzees and gorillas. They are also able to put themselves in the position of others, recognize causalities, and draw conclusions. Pigeons can learn English spelling up to the level of six-year-old children. Behavioral studies have shown that these birds can fashion and use tools, and recognize people on the street, things that even many primates fail to accomplish.

Some octopus species have been documented digging for and using seashells and coconut shells as tools and protection while other species have collected rocks and positioned them in front of dens as a way to safeguard them (FIG 2.8). There are many anecdotal stories of octopuses escaping from tanks in aquariums and shooting jets of water at particular individuals and equipment. This may sound more entertaining than indicative of intelligence but the stories also demonstrate that the animals can *recognize* individual humans and show an element of planning and evaluation of their surroundings.

Octopuses lack a central brain. Each of an octopus's eight arms has an extensive number of neurons resulting in the equivalent of having a "brain" in each appendage that is capable of receiving and processing information about the environment. Instead of a central nervous system, their "brain" is their body and their body is their brain. These findings question a clear-cut link between brain size and cognitive skills. This brings us to the next subject.

In humans, what happens when the brain is critically impaired or in large part missing? Radical removal of half of the brain is sometimes performed as a treatment for epilepsy in children. Commenting on a cohort of more than fifty patients who underwent this procedure, a team at Johns Hopkins University in Baltimore wrote that they were "awed by the apparent retention of memory after removal of half of the brain, either half, and by the retention of the child's personality and sense of humor."

Now consider the following case from China of a twenty-four-year-old woman admitted to the PLA General Hospital in Shandong Province complaining of dizziness and nausea. She told doctors that her speech only became intelligible at the age of six and that she hadn't walked until she was seven.

A CAT scan—which uses computerized axial tomography to produce cross-sectional images of the body—immediately identified the source of the problem: her entire cerebellum was missing.[1] The space where it should have been was empty of tissue. Instead, it was filled with cerebrospinal fluid. The patient's doctors suggested that normal cerebellar function might have been taken over by the cortex but was it? We shall revisit this "explanation" shortly.

The medical literature contains a surprising number of known cases of people missing a substantial portion of their cerebral cortex, the outermost layer of brain tissue, held to be the seat of our thinking brain. A currently living and healthy ten-year-old German girl is one. She was born without the right hemisphere of her cortex, though this wasn't discovered until she was three years old. According to Lars Muckli from the University of Glasgow, who led the study despite the fact that his patient was lacking one hemisphere, she demonstrated normal psychological functioning, and managed to live a perfectly regular and fulfilling life. He described her as witty, charming, and intelligent.

In an article titled "Is Your Brain Really Necessary?" science writer Roger Lewin reviewed a series of six hundred cases by English pediatrician John Lorber of people with *hydrocephalus*—an excess of cerebrospinal fluid, commonly known as water on the brain. In sixty of those cases the fluid took up 95 percent of their cranium (skull), and yet, half of those had above average IQs. Among them was a student with an IQ of 126 who received a first-class honors degree in mathematics and was deemed socially normal. For this case, Lorber noted that instead of the typical 4.5 cm thickness of brain tissue between the ventricles and the cortical surface, there was just a thin layer of mantle measuring a millimeter or so. The cranium was mainly filled with cerebrospinal fluid.

In July 2007, a forty-four-year-old Frenchman went to a hospital complaining of a mild weakness in his left leg. When doctors learned that the man had a spinal shunt removed when he was fourteen, they performed a *computed tomography* (CT) scan and *magnetic resonance imaging* (MRI) scan. What they discovered was a huge fluid-filled chamber occupying most of the space in his skull, leaving little more than a thin sheet of actual brain tissue (like figure 9). While the brain was virtually absent, intelligence tests showed the patient to have an IQ of 75 (the average score is 100). Today, this would be considered borderline intellectual functioning.[2]

The patient was a married father of two, and worked as a civil servant apparently leading a normal life, despite having enormously enlarged ventricles with a decreased volume of brain tissue. "What I find amazing to this day is how the brain can deal with something which you think should not be compatible with life," commented Dr. Max Muenke, a pediatric brain defect specialist at the National Human Genome Research Institute.

Neuroscientists explain the near normal behavior of people with hydrocephalus on the basis of *neuroplasticity*. Proposed mechanisms include neurogenesis (creation of new neurons), programmed cell death (a questionable premise), and experience-dependent synaptic formation.

The argument for this counterintuitive view rests precariously on the proposition that only the presence of unimaginably high levels of "redundancy" or "plasticity," obviously absent, could make up for the drastic reduction in brain mass in certain, clinically normal hydrocephalic cases or in people who have had a large part of their brains removed.

Another common explanation is based on the finding that while the damage to the hemispheres is typically extensive, the children's or later the adults' *brain stems* are usually (but not always) unaffected and assume the functions of the missing brain tissue.

I don't think these hypotheses explain satisfactorily how a very thin sheet of cerebral cortex can function in the absence of large parts of the brain. How can a greatly reduced number of neurons and glial cells function as well as the full complement of cells in a normal brain? Imagine, for example, that you broke your right leg. Even with good healing, no amount of stimulation, physiotherapy, or exercise will return your leg to its former glory. It seems to me that words like *neuroplasticity* or the hypothesis that the brain stem assumes extra functions are a smoke screen for lack of knowledge. Logic dictates that there must be a limit to how much a so-called unaffected area in the brain or a broken tibia (one of two leg bones) can compensate for lost tissue.

And as far as unaffected or healthy parts of the brain are concerned, which are supposed to acquire the function of the lost parts, I wonder—how could there be any areas in the brain not affected by loss of surrounding tissue and the pressure exerted on them by the cerebrospinal fluid? That is how brain tumors gradually destroy brain tissue and function.

I suggest that people with missing brain matter who appear to act quite normally perform as well as they do not because of "neuroplasticity" or "recruitment" of unaffected areas in the brain, though no doubt some of that applies, but because the brain never works alone. Its function is inextricably linked to the body and to the outside world. In the individual who lacks a large part of their cortex, the neurons in the cranial nerves, spinal cord, and other cells in the body (somatic cells, immune cells, heart cells, etc.) form a network that is constantly communicating with the brain—or what's left of it—and acts almost like a backup disk on a computer, containing snippets of memory and functionality that collectively contribute to near normal cognition and behavior.

An Abundance of Neurons

Let's now move to a subject that is almost diametrically opposed to lack of brain tisssue; namely, *savant syndrome*. Just before midnight, in 2002, as Jason Padgett was leaving a Tacoma karaoke bar, two guys jumped him from behind. At the emergency room, doctors told Jason that he had suffered a severe concussion and had a bleeding kidney. He was given painkillers by the attending staff and sent home. In my estimation, a rather cavalier response by the doctors to a potentially life-threatening condition. But let's proceed.

Soon after the mugging, Jason started experiencing the world differently. He saw everyday objects as geometric patterns and would talk incessantly about math, pi, and infinity. Jason also had an urge to draw. He began drawing complex, fascinating figures using only a pencil and a ruler. He said he had no idea what he was drawing.

Eventually, Jason consulted the world's foremost authority on savant syndrome, Darold Treffert. Savant syndrome is usually described as islands of genius and ability in persons who clearly "know things they never learned." Some are born savants like Mozart; others are made, like Jason. Skills most often exist in art, music, calculating, and mechanical or spatial abilities. Whatever the special skill, it is always associated with massive, exceedingly deep memory but very narrow within the area of the special skill.

Treffert, a Wisconsin psychiatrist, had studied savant syndrome for over fifty years. He is of the opinion that the brain of a person who has suffered a severe head injury is capable of "recruiting" another part of the brain to compensate for the part that was damaged. In Jason's case, he suggested the portion capable of doing high-level math was recruited. Fancy word—*recruiting*. Does it really explain a process that involves epic intellectual transformation?

Since most savants are born with special gifts and often with cognitive challenges such as autism while a few develop exceptional expertise following some physical trauma to the head, it must be assumed that savantism is genetic. In some people that exceptional talent is apparent early in life, in others it remains dormant until triggered by an environmental event, such as a blow to the head.

Michael S. Gazzaniga is a professor of psychology at the University of California. In his book, *The Mind's Past*, he writes, "As soon as the brain is built, it starts to express what it knows, what it comes with from the factory. And the brain comes loaded." Savants are examples of the fact that we do not start life with a blank slate. (Freud called it tabula rasa.) Just the opposite.

However, there is a problem with accepting the genetic transmission of special intellectual gifts as a full explanation of savantism. It would be adequate if we could prove that all savants had brilliant parents or grandparents or great-grandparents, and there is no proof of that. So where did these "genius genes" come from? The answer seems to be: they were created either by a spontaneous mutation (there is no evidence for that either) or all of us have these genes lying dormant but they only get activated by epigenetic mechanisms. The challenge is how to release this latent capacity nonviolently, without a brain injury or similar traumatic event.

Surely, the proclivity, the extraordinary ability must always have been present but in an inactive state. You might say it was locked away and the injury, in Jason's case, unlocked it. I think the simplest, most logical, and scientifically sound explanation for savant syndrome is on the basis of epigenetics. The genes that control these special musical, mathematical, or other exceptional talents were expressed in the born savants whereas in the late bloomers these same genes were switched on by environmental events like a blow on the head. Of course, because we don't yet have savant mice or rats, it will take some time before this hypothesis can be proven in the laboratory.

The Good News

We have known for some time that omega-3 and omega-6 fatty acids promote healthy aging of the brain. However, findings from two novel studies support a critical role for polyunsaturated fatty acids in maintaining intelligence and memory. These two studies stress the importance of investigating the effects of a class of nutrients jointly rather than one at a time. By reducing inflammation, oxidative stress, and platelet accumulation, as well as improving blood pressure, fatty acids as a group have physiological effects that can improve brain health. Foods rich in these acids are fish, nuts, seeds, and oils, essentially the so-called Mediterranean diet.[3]

And here comes more praise for the Mediterranean diet. Domenico Praticò of Temple University has shown that the consumption of extra-virgin olive oil protects memory and learning ability and reduces the formation of amyloid-beta plaques and neurofibrillary tangles in the brain—classic markers of Alzheimer's disease. "We found that when it detects it olive oil reduces brain inflammation but most important activates a process known as *autophagy*." Autophagy is the means by which cells break down and clear out intracellular debris and toxins, such as amyloid plaques and tau tangles.

Nowadays, worry about memory loss with aging seems to be on everyone's mind, young and old. A study recently published by Jason Steffener, a scientist at Concordia University in Montréal, has thrown new light on this subject by showing that the more flights of stairs a person climbs, and the more years of school a person completes, the "younger" their brain appears physically.[4] The researchers found that brain age decreases by 0.58 years for every daily flight of stairs climbed between two consecutive floors in a building and by 0.95 years for each year of education.

You probably knew all along but now there is scientific proof: Green spaces are good for gray matter. A study at the universities of Edinburgh and York aimed to understand how older people experience different urban environments using electroencephalography (EEG), self-reported measures, and interviews. As part of the experiment, eight volunteers aged sixty-five and over wore a mobile EEG headset, which recorded their brain activity when walking between busy and green urban spaces. The research team also ran a video of the routes the people walked, asking the participants to describe "snapshots" of how they felt. The volunteers were also interviewed before and after their walks.

Older participants experienced beneficial effects of gardens, lawns, and trees while walking in busy urban environments. The presence of green spaces promoted personal and cultural memories and social connections.

Summary

The most recent discoveries in the realm of neuroscience at least question and at most refute the long-held theory that memory is stored in the synapses of the brain. The embodied brain hypothesis postulates that new information and experiences trigger the formation of engrams with enduring physical or chemical changes in neurons and in every cell in the body. Neurons make up only 15 percent of our brain cells. Glial cells, dismissed for years as mere connective tissue and otherwise of little importance actually control to some degree of communication between neurons and play a central role in learning. Astrocytes, in particular, significantly affect how information is transmitted and stored in the brain. Furthermore, as Jorge Colombo pointed out, astrocyte networks may serve as nonneuronal channels of information processing, in which glia cells couple neurons and synapses into functional assemblies.

Scientists have demonstrated that neurons in the spinal cord are able to learn motor skills independently of the brain. Similarly, the enteric nervous system can act independently of the brain but at the same time is intimately connected to it. Evidently, neurons outside of the brain, not just in the brain, retain memories.

People who have had parts of their brain tissues surgically removed or were born hydrocephalic perform as well as they do, not because of "neuroplasticity" or "recruitment" of unaffected areas in the brain but because the brain, or whatever is left of it, is part of a much larger communication network in the body. This extended brain, the embodied brain acts as a backup memory storage system for the enskulled brain, and when it detects missing data "above" sends the required information there.

In the next chapters we shall discover how other cells in the body (immune cells, somatic cells, and cells in organs such as the heart) contribute to the dynamic system that is the embodied brain.

KEY TAKEAWAYS

- Recent research subverts the hegemony of the synapse.
- The synapse acts as a conduit to neurons not as a depository for memories.
- Though it is agreed that neurons in the cerebral cortex store memories, other cells such as glial cells and neurons in the spinal cord and the gut also contribute to information processing and storing.
- It is time for responsible neuroscientists to seriously consider modifying or abandoning the cortico-central hypothesis of memory.
- Cases of hydrocephaly provide convincing support for the argument that the size of a human brain is unrelated to its information content, intelligence, or capacities.
- Savants are born with extraordinary abilities. The genes that control these exceptional talents were active in the early childhood savants whereas in the late bloomers these same genes were switched on by environmental events, like a blow to the head.
- Many factors, including genetic endowment and physical and mental activity, influence cognition and brain aging, but nutrition plays an essential role.

CHAPTER THREE

THE IMMUNE SYSTEM: THIS DOCTOR MAKES HOUSE CALLS

Introduction

Our bodies are patrolled 24/7 by a silent army of vigilant police officers, health inspectors, doctors, and ambulances. They protect us from uninvited guests, including bacteria, viruses, and fungi, as well as intrusive substances, such as a splinter or an ingrown toenail. Basically, anything that does not belong in our bodies attracts their attention. They repair injuries and dispose of waste. And they keep records of the miscreants they encounter for future reference. Our immune system does all this and more to keep us healthy.

How It Works

The immune system has evolved to recognize and respond to threats to health, and to provide lifelong memory designed to prevent recurrent disease. The immune system is made up of a large variety of white blood cells, some of which circulate through our arteries and veins, while others reside in various tissues of the body, including the lymph nodes and the skin (FIG 3.1).

The immune system protects the body by producing antibodies, which are proteins that help to prevent intruders, called antigens, from causing much harm. After an infection with a pathogen, a cascade of reactions will usually be set into motion. The immune system will start to produce the specific type of antibody designed to protect against that particular pathogen, among others. The body is capable of creating literally billions of antibodies that are adept at fighting off billions of potential invaders.[1] And our immune cells are very clever. They can learn on the job, remember what they have learned, and can apply this education in response to future challenges.

Among the key players in the immune system are two types of white blood cells, known as *T cells* (FIG 3.2) and *B cells*. B cells make antibodies specific to particular pathogens. T cells change their own shape in order to surround, engulf, and thus destroy the invader. Then, after the disease is eliminated some T cells and B cells are converted into *memory T* and *memory B cells*.[2]

Vaccination (immunization) is based on this premise. When a person is immunized (for polio, measles, etc.) they are given a small dose of the appropriate antigen, such as dead or weakened live bacteria, in order to activate immune system "memory," which then allows the body to react quickly and efficiently to future exposures. This is also why, generally speaking, once you've had chicken pox or some other childhood illness, you don't catch it again—because the antibodies that were activated by your getting it the first time are still there in your immune system.

An interesting corollary to this kind of learned immunity is something called the Hoskins Effect[3] (FIG 3.3), which refers to the fact that the immune system responds most powerfully when it reencounters the exact same infection to which it was previously exposed, and less powerfully to a slightly different strain or version of it. This explains why the annual flu shot may work better some years than others or be more effective against one strain of flu than another.

Cells typically get most of their energy from glucose and other sugars. When those fuels run low and oxygen is still available, memory T cells are able to enhance their own survival by packing themselves full of mitochondria, which are a cell's energy generators. I shall discuss mitochondria in greater detail in the next chapter. Mitochondria allow cells to produce energy efficiently from alternative fuel sources such as fats and amino acids. Consequently, these cells can live a long time, which means that their memories persevere for a long time.

Peter Cockerill, from the University of Birmingham, demonstrated that a single cycle of activation of *naïve T cells* leads to long-term epigenetic changes in these cells. Cockerill proposed that this forms the basis of a **long-term memory** that allows for an immediate response when the body encounters an infection and T cells are activated for a second time. Thomas Dörner, from the University of Berlin, when writing on the subject of antibodies speaks of *long-lived plasma cells*[4] secreting specific antibodies even in the absence of a virus and suggests that this is how they protect against reinfection. In his opinion, these secreted protective antibodies of **humoral memory** provide an efficient line of defense against reinfection and are backed up by specific B and T memory cells of **reactive memory**.

And there is more fascinating news about the immune system from the University of Vienna about NK cells in the liver, the organ that is generally assumed to be a large reservoir for NK cells. NK cells are natural cytotoxic killer cells in human blood and are a type of lymphocyte, a subgroup of white blood cells. Up until now, NK cells were regarded as having no memory function, meaning that they are unable to kill on an "antigen-specific" basis but are only able to react afresh each time to viruses and sources of infection in a nonspecific way.

In addition to the memory cells that are on patrol in the circulation, Australian scientists have discovered that the immune system also leaves behind a garrison of memory cells, *follicular memory T cells*, strategically positioned at the entrance of the lymph nodes, particularly those that are potential sites of microbe reinvasion, like in the neck, under the armpits, and in the groin area, to screen for return of antigens they have encountered before. This is an important finding because, until now, immunologists have thought that memory is provided by only circulating cells.

And speaking of lymph nodes, Jochen Hühn at the Helmholtz Centre for Infection Research in Germany has pointed out that lymph nodes are basically the immune system's meeting points. His team also found that the location of the lymph nodes was directing the development of the immune cells they contained. They did so by collecting lymph nodes from various parts of rats' bodies and transplanting them to different locations. It seems that the transplanted cells retained their original capacities for weeks. Since all the cells within the lymph node, including immune cells, are constantly regenerated, the researchers further concluded that this memory was being encoded and passed on from one generation to the next.

Now, research led by a team from the University of Pennsylvania School of Veterinary Medicine found that, after infection with the parasitic disease leishmaniasis, a population of T cells with a memory for the parasite remained in the skin. This was actually the first time that T memory cells were found to reside in the skin, and the discovery holds great promise for immunization against tissue-specific diseases based on a skin-scratching process called scarification that was formerly used for smallpox. Scarification has been shown to effectively generate *tissue-resident memory cells* and may be used in the future to protect people from leishmaniasis.

Turning to the brain now, it has been well established that under healthy conditions, immune cells reside in the central nervous system. These include microglia, dendritic cells, and mast cells, along with T cells and B cells. Astrocytes are derived from neural stem cells, but they also secrete and respond to inflammatory mediators, so astrocytes are also included in this system. When activated by a pathogen or stress hormones microglia secrete both proinflammatory cytokines and chemokines to recruit other immune cells to sites of damage, and anti-inflammatory cytokines and growth factors to stimulate regrowth and repair within the central nervous system. Microglia are capable of orchestrating a potent inflammatory response. They are also involved in synaptic organization, trophic neuronal support during development, phagocytosis of dead cells in the developing brain, myelin turnover, control of neuronal excitability, phagocytic debris removal, as well as brain protection and repair. You wonder, is there anything microglia don't do?

Jonathan Kipnis, director of the University of Virginia's Center for Brain Immunology and Glia, created a scientific breakthrough in 2014 with his discovery of meningeal lymphatic vessels. The meninges are membranes that envelop the brain and spinal cord. Lymphatic vessels are thin tubes, constructed like blood vessels that transport lymph (a clear fluid that carries infection-fighting white blood cells) throughout the body. Meningeal lymphatic vessels directly link the central nervous system (CNS), including the brain, to the lymphatic system, and their discovery overturned decades of textbook teaching that the brain is "immune privileged"—in other words, lacking a direct connection to the immune system—and galvanized the scientific world to seriously investigate **how the complex bidirectional communication between the immune system and the brain affects health and disease.** Multiple new lines of evidence support the notion that immune cells are actually required in the brain for optimal neuronal survival following CNS injury.

Exposure to inflammation caused by a viral infection during pregnancy has been linked to an increased likelihood of conditions with altered neuron development, such as autism and schizophrenia. Recent research at King's College London has provided us now with new explanations of this phenomenon. It seems that during a viral infection, inflammatory cytokines are produced as part of the immune response and it is these proteins that set off symptoms such as fever, headache, etc. The research focused on the impact that *interferon gamma* (of one of these proteins) has on neural progenitor cells which are the cells present in the developing brain that eventually become the adult brain cells or neurons. The researchers demonstrated that the cellular memory of the interferon gamma remains intact as neural progenitor cells mature into brain cells, producing a series of molecular and genetic changes thought to contribute to developmental disorders of the brain such as autism or schizophrenia.

Deepak Srivastava, joint senior author of the research, stated, **"Our study has identified a possible way in which this cellular memory of one component of a maternal viral infection is embedded at a molecular and genetic level in the developing brain."** This study tells us that the immune system does much more than fight infections. It actually affects how neurons develop.

In the last decade there is abundant evidence in the medical literature that a healthy immune system supports normal brain function, while an altered one results in cognitive impairment. Furthermore, without the normal homeostatic functions of brain microglia, the ability to cope with pathology is likely hindered. A failing immune system and an ongoing disease that increases demand for immune cells further fuels the pathology. Neuroscientists from the University of Virginia, Charlottesville, have suggested that a decrease in the number or a malfunction of T cells during gestation leads to a dysregulation of synapses, neural circuits, and ultimately contributes to the development of autism spectrum disorder.

Robert H. Yolken and his team at the Johns Hopkins University School of Medicine focused on research of mania. They observed that during manic episodes, many patients register elevated levels of cytokines, molecules secreted by immune cells. He prescribed thirty-three mania patients who had previously been diagnosed with a manic episode a probiotic prophylactically. These patients proved 75 percent less likely to develop another manic attack compared with patients who did not take the probiotic. The study is preliminary, but it suggests that targeting immune function may improve mental health outcomes and that supporting the microbiome (we shall discuss the microbiome in detail in the next chapter), might be a practical, cost-effective way to do this.

At least a half million women in the United States each year suffer from postpartum depression, and that is probably a low estimate. It is surprising how little we know about this very serious disease. Previous research has focused primarily on potential hormonal etiology, though some earlier work has been done on the immune system. In those studies, scientists have looked at signs of inflammation in the blood and found mixed results.

A study at Ohio State University took a different approach. They stressed rats during pregnancy. The stressed animals exhibited behaviors similar to those exhibited by mothers suffering of postpartum depression; that is, decreased attentiveness to their pups, depression, and anxiety. Unlike unstressed comparison animals, the stressed rats had higher levels of inflammatory markers in their brain tissue. Furthermore, the researchers found evidence that stress leads to changes in how microglia function.

Significantly, the researchers failed to observe evidence of increased inflammation in the blood but detected signs of inflammation at the medial prefrontal cortex, a mood-related brain region previously implicated in postpartum depression. Study coauthor Kathryn Lenz said that outcomes might improve by treating brain inflammation and the reaction of the immune system to the inflammation in addition to the present emphasis on medication, diet, and stress reduction.

What role does the immune system play in how we respond to stress? We know that stress increases the production of cortisone by way of the HPA axis. Increased cortisone levels inhibit the immune system. That is why when people are under stress or feel anxious or depressed, they are more likely to suffer a cold or the flu or have higher rates of infections. On the other hand, when the adrenal glands become depleted and can no longer produce adequate amounts of cortisone, the immune system goes into overdrive and may turn against its own body. Ideally, what you want is a Goldilocks condition[5]—not too much and not too little cortisone—just the right amount. Biologists refer to a system that is in equilibrium as being in a state of *homeostasis*.

The Good News

How do we counteract the effects of stress to our existence at the molecular level? What are the concrete things we can do to actively promote more favorable gene expression, particularly in the immune system? One answer is through mind-body practices like meditation, which, according to Steven Cole (who we have referred to before), has been shown to cultivate positive and happy immune cells. Research has linked meditation to higher antibody production, reduced negative inflammatory activity, increased positive antiviral response, and improved function of specific strains of immune cells.

In recent years, a new field of study, known as mind-body genomics, has emerged. Through a series of studies, Nobel laureate Elizabeth Blackburn, a biochemist at the University of California, San Francisco, and her colleague, psychiatrist Elissa Epel, found that meditation could affect the ends of DNA known as the *telomeres*, which act as protective caps for genes. Once a cell's DNA loses its telomere it can no longer divide, and dies. Massive cell death brings about aging. Telomeres, like immune cells, respond to emotions. Negative emotions shorten telomeres, while happy feelings help to maintain their structure.

More than a century ago, scientists demonstrated that sleep supports the retention of memories of facts and events. Recent studies have shown that slow-wave sleep, often referred to as deep sleep, is important for transforming fragile, fresh memories into stable, long-term memories. Now, researchers propose that deep sleep may also strengthen immunological memories of previously encountered pathogens. Therefore, getting plenty of deep sleep contributes significantly to overall good health.

Zinc plays an important role in immunity. Low or deficient levels of zinc profoundly affect the number of your immune cells that are available to fight an invader. After examining thirteen randomized placebo-controlled studies between zinc and the common cold, researchers found that taking zinc within twenty-four hours of the first signs of a cold could shorten its duration and make the symptoms less severe.

Garlic has been used by many different cultures to fight infectious diseases, as it has antimicrobial, antifungal, and antiviral properties. Modern studies have shown that incorporating garlic into one's diet may actually activate immunity genes and reduce the severity and incidence of a cold.

Summary

The memory B cells, memory T cells, and follicular memory T cells of the immune system play a crucial role in protecting our bodies from potentially harmful invaders. Brain-resident immune cells, particularly microglia, play a key role in protecting healthy functioning of the central nervous system. Furthermore, lymph nodes have a **location-specific memory** that is assumed to be encoded in its *stroma*; that is, in its cellular membrane. These cells of the immune system develop into a long-lasting "memory population" that will protect the body against future infections. All evidence points to the existence of a complex feedback loop between the immune system and the brain.

This is all straightforward, totally unchallenged, universally accepted academic science. Yet, old-school neuroscientists still cling to the belief that only cortical (brain) neurons and their synapses cache memories. The way the immune system operates contradicts that view. In addition to neurons, the four types of immune cells referred to here are also capable of both storing and sharing information/memory with each other.

The immune system, with its overabundance of memory cells and geographical memory lymph nodes, is a microcosm of the embodied brain: unseen, silent yet constantly affecting the embodied mind while short-circuiting consciousness. Like the neurons we spoke of in the previous chapter, immune cells are capable of both storing and sharing information/memory with each other and the rest of the embodied brain. It's like a dinner party with a group of people. Everything that one person says is heard and reacted to by all those present.

Memory cells of the immune system collect a treasure trove of information on the predators they encounter over a lifetime. Does it not make sense to at least consider the possibility that cells as clever as these immune cells might not, like neurons, also contain fragments of life memories? I hope the next chapter on the cell will shine more light on this question.

KEY TAKEAWAYS

- At the first hint of invading organisms the immune system launches an orchestrated attack, usually destroying and eliminating them.
- Our immune system will memorize the pathogen after an infection and following reinfection recognize and destroy the same pathogen.
- A total of three different cells in the immune system retain memory. They are memory B cells, memory T cells, and follicular memory T cells.
- Immune cells respond to emotions.
- Deep sleep enhances memory formation in the brain and in the memory cells of the immune system.
- A well-regulated immune system is necessary for normal brain function.

CHAPTER FOUR

MYSTERIES OF THE HUMAN CELL

Introduction

An adult human body contains about 50–100 trillion cells. (That's 100,000,000,000,000!) These cells are constantly in a state of flux; some are created while others are destroyed. About 300 million cells die every minute in our bodies! The only cell visible to us without the use of a microscope is the ovum. In spite of their minuscule size, these tiny biological machines are surprisingly intelligent.

We do not generally think of somatic cells as intelligent or storing data other than those relevant to their function. However, when these cells form networks, as they do in our bodies, their collective intelligence emerges. Those capacities are genetically encoded into every living cell. In that sense, our cells are comparable to computers programmed with software referred to as artificial intelligence. Just like AI-equipped computers change and learn from feedback, so do our cells.

Some biologists have started to apply the concepts of electrical engineering to living cells, literally programming them for data storage and computation. The findings of these truly revolutionary studies are contributing to our understanding of how biological cells archive information, work in assemblies similar to neural networks, and how they connect to the latter.

The Cellular Universe

There are over two hundred different types of cells in our bodies, displaying a variety of sizes from small red blood cells that measure 0.00076 mm to liver cells that may be ten times larger. Different cells serve different purposes. **About ten thousand average-sized human cells can fit on the head of a pin.** In mammalian cells approximately two microns of linear DNA have to be packed into a nucleus of roughly ten μm diameter (ten μm = the width of cotton fiber).

Actually, there are two types of DNA: the so-called *nuclear DNA* that we inherit from both parents and *mitochondrial DNA* (a small amount of DNA that we carry in the mitochondria; see below) that we inherit only from our mothers. The reason for that is that during fertilization a father's sperm destroys its own mitochondrial DNA. The same physiologic processes carried out in our bodies by organs such as lungs, heart, kidneys, etc., are performed in our cells by thirty-seven diminutive organ systems called *organelles* (FIG 4.1).

Two of the most important organelles are the *mitochondrion* and the *cytoskeleton*. Mitochondria regulate cell survival and metabolism. They are often called the powerhouse of the cell because they take in oxygen and nutrients, break them down, and create energy-rich molecules for the cell. This is essential for cells and tissues to operate properly and defects in mitochondrial functions and number are linked to aging and chronic diseases such as cancer, obesity, type II diabetes, and neurological disorders.

In addition to supplying cellular energy, mitochondria control cellular differentiation, the cell cycle, and cell growth. Lately, mitochondria have been increasingly recognized as playing an important role in the aging process. Researchers at Boston University have identified a protein, *G-Protein*

Pathway Suppressor 2 (GPS2), that moves from a cell's mitochondria to its nucleus in response to stress and during the differentiation of fat cells. This finding demonstrates that there is communication from at least one organelle to the nucleus and not, as was believed in the past, just from the nucleus to the rest of the cell. Thus, there is much communication within the cell and, as we shall shortly see, also between cells.

The cytoskeleton consists of a cellular scaffolding or skeleton within a cell's cytoplasm. It is made up of *microtubules*, assembled from the protein *tubulin* into tube-shaped networks. Some studies support the view that it is the microtubules within the cytoskeleton that store memories. Nancy Woolf at the University of California, Los Angeles, concluded in the book *The Emerging Physics of Consciousness* that there are links among microtubules, memory, and consciousness. This will be revisited in the last chapter.

The cytoskeleton is one of the best candidates for mechanisms underlying information processing at the single-cell level. The cytoskeleton has all of the necessary properties: it is a large, complex organelle that is readily modified by a variety of molecular pathways (writing data), is interpreted by numerous motor proteins and other machinery (reading data), and implements a rich set of discrete transition states that could implement computational operations.

Other organelles are *lysosomes*, which are critical garbage disposal and recycling centers, *flagellum*, a lash-like appendage that protrudes from the cell body (all mammalian sperm cells propel themselves using their flagellum), and many others. Virtually all of the cell's genes are contained within the cell's largest organelle, the *nucleus*.

In addition to organelles cells contain proteins. Proteins, which are products of genes, are the workhorses of cells. When the cell receives a signal, such as to *grow*, *divide*, or *change shape*, new proteins are manufactured or old ones are moved around within the cell to accomplish this task. The Donnelly Centre of the University of Toronto has constructed maps of about three thousand proteins in yeast cells. Human cells perform tasks similar to yeast cells except that they have four times the yeast cells' number of proteins. That means that a human cell contains **twelve thousand proteins** that amounts to **forty-two million protein molecules** in a space so tiny it is invisible to the naked eye.

How could so much "stuff" be inside something no larger than a grain of sand? The following analogy may prove helpful in this respect. To an observer from far away in the universe our planet would appear as this tiny speck of dust in a galaxy surrounded by other galaxies stretching into infinity. This astronomer could not possibly conceive of the millions of plants, animals, and intelligent life-forms existing in such a minuscule space. Yet it follows from the evidence presented here that there is at least as much room, diversity of elements, and capacity for memory storage within a biological cell as there is in our world. What one sees very much depends on one's vantage point and acuity of vision.

And now, let's take this vantage point idea a bit further. Let us consider for a moment the atom. An *atom* is the smallest constituent unit of ordinary matter that has the properties of a chemical element. Atoms measure around one hundred picometers, which equals one ten-billionth of a meter. Within this atom, there are the old standbys: protons, neutrons, and electrons, and the new breed of subatomic particles with weird names like *muon*—a fat, short-lived cousin of the electron—*quarks*, *gluons*, and recent arrivals called *pions*, *kaons*, *taus*, *lambdas*, *sigmas*, and *xis*. There are matter particles and force particles with masses ranging from zero (*photons* and *gluons*) and near zero (*neutrinos*) to the *top quark*, which is as hefty as an entire atom of tungsten—an element whose name, in Swedish, means "heavy stone." Compared to atoms, biological cells are huge.

The Language of Cells

For a long time scientists have wondered how cells sense light, heat, propagate nerve signals, or react to changes in the environment. Physicists from Oregon State University and Purdue University have shown that when cells meet, a small channel usually forms between them called a gap junction.

On an individual level, a cell in response to *adenosine triphosphate*[1] begins to oscillate, part of its call to action. With gap junction-mediated communications, most of the cells eventually decide what the correct sensory input is, and the signal that gets passed along is pretty accurate. "The thing is, individual cells don't always get the message right, their sensory process can be noisy, confusing, and they make mistakes," Bo Sun, an assistant professor of physics at Oregon State University, said. "But there's strength in numbers, and the collective sensory ability of many cells working together usually comes up with the right answer. This **collective communication** is essential to life."

There are many times when our cells need to move. Immune cells hunt for unwanted intruders. And healing cells (*fibroblasts*) migrate to mend wounds. Of course, not all movement is desirable: when cancer cells metastasize, trouble follows. When moving, the cell converts chemical energy into mechanical force. This force can now be measured.

In a study at Sanford Burnham Prebys Medical Discovery Institute and the biosensors team at the University of North Carolina at Chapel Hill, scientists used the cryo-electron microscope, artificial intelligence, and tailor-made computational and cell imaging approaches to compare nanoscale images of mouse fibroblasts to time-stamped light images of fluorescent *Rac1*, a protein that regulates cell movement, response to force or strain, and pathogen invasion. This technically complex workflow—which bridged five orders of magnitude in scale (tens of microns to nanometers)—took years to develop to its current level of robustness and accuracy. The researchers concluded that when cells are not traveling, they do not just work side by side with their neighbors without "talking" to them. Rather, they form contacts with each other: connections of different size, strength, and duration.

One of the means by which cells communicate with each other is by *trafficking vesicles*—"transport pods" that sprout from one cellular compartment and fuse with another. These vesicles move molecules between different compartments within a cell and between cells. They carry material to the *Golgi apparatus* and to the *endoplasmic reticulum*.[2]

Cell-to-cell communication is also facilitated by the formation of pores, referred to as gap junctions, between physically attached cells as well as by the secretion of signaling molecules. These consist largely of extracellular RNA (exRNA). RNA molecules were thought to only exist inside of cells, but now they have been found to also occur outside of cells and participate in a cell-to-cell communication system that delivers messages throughout the body.

Recently, in addition to these signaling molecules, a novel mechanism was discovered: *membrane nanotubes* or *tunneling nanotubes (TNTs).* TNTs are long, thin translucent filaments about 50 nanometers wide and 150–200 microns long, extending between cells. They act as conduits for sharing miRNAs, messenger RNAs, proteins, viruses, and even whole organelles, such as lysosomes and mitochondria.

Healthy adult cells don't usually make TNTs, unless stressed or sick. Troubled cells send chemical SOS signals to neighboring healthy cells. In response, healthy cells extend TNTs through which they transfer healing substances, proteins, and miRNAs to their ailing neighbor (FIG 4.2). Who would have thought of human cells as Good Samaritans? Yet this is the case and is further clear evidence of the intelligence of cells.

About ten years ago, scientists began discovering a new communication system between cells that is mediated by *extracellular RNA (exRNA).* Researchers from Baylor College of Medicine detected six major types of exRNA cargo in bodily fluids, including serum, plasma, cerebrospinal fluid, saliva, and urine. The system seems to work both in normal as well as diseased states.

Still on the subject of cell-to-cell communication, researchers from the Riken Institute in Tokyo have systematically analyzed the relationship between *ligands*, substances such as insulin and interferon that carry messages between cells, and *receptors*, the proteins on cell surfaces that receive these messages. They succeeded in elucidating and quantifying the repertoire of signaling routes between different cell types. Based on the analysis, the authors gained new insights into how cells communicate. According to Jordan Ramilowski, the first author of the study, "One intriguing conclusion is that signals between cells of the same type are surprisingly common."

The cells in our bodies can divide as often as once every twenty-four hours, creating new identical copies of themselves. DNA binding proteins called transcription factors (ethyl and methyl groups that we called simply *switches* in chapter 1) are required for maintaining cell identity. They ensure that daughter cells have the same function as their mother cell, so that for example liver cells can make liver enzymes or thyroid cells can produce thyroid hormone. Each cell type can be distinguished based on its epigenome.

During each cell division, the transcription factors are removed from DNA and must find their way back to the right spot in the newborn cell. Jussi Taipale, professor at the Karolinska Institute and the University of Helsinki, and his research team, have discovered a possible mechanism for how the transcription factors find their correct places. It seems that a large protein complex called *cohesin* encircles the DNA strand as a ring does a finger, thereby helping the transcription factors to find their original binding region on both DNA strands.

"Now we have found a possible mechanism for how this **cellular memory** works, and how it helps the **cell remember** the order that existed before the cell divided," explained Jussi Taipale. His words, not mine.

Using a micropipette adhesion frequency assay, scientists at Georgia Tech and Emory University have observed that molecular interactions on cell surfaces may have a **memory** that affects their future interactions. Up to now, researchers who used sequentially repeated tests to obtain statistical samples of molecular properties usually assumed that each test is identical to and independent of any other tests in the sequence. Cheng Zhu, the lead investigator, revealed examples in which an interaction observed in one test affected the outcome of the next test. Depending on the biological system, the effect could either increase or decrease the likelihood of a future interaction.

Supported by the National Institutes of Health, the research demonstrates that certain cells can **remember their earlier encounters** through specific receptor-ligand interactions. Zhu said, "This may represent a way for cells to regulate their adhesion and signaling. For T cells, the ability to 'remember' even a brief interaction with a pathogen may be related to their ability to tell an intruder from 'self' molecules, which is crucial to the body's defense in the immune system."

Amit Pathak, assistant professor of mechanical engineering and materials science at Washington University in St. Louis, created a device that can measure how long a cell's memory lasts when transferred from one environment to another. He found that cells remember the properties they had in their first environment for several days after they move to another in a process called *mechanical memory*.

The discovery by Ramesh Shivdasani, at the Harvard Stem Cell Institute, went even further back in time. He found that **adult tissues retain a memory, inscribed on their DNA, of the embryonic cells from which they arose**. The study led to an even more intriguing finding, that the **memory is fully retrievable**. Cells can play the story of their development in reverse to switch on genes that were active in the fetal state. Said Shivdasani, "Beyond the sheer existence of this archive, we were surprised to find that it doesn't remain permanently locked away but can be accessed by cells under certain conditions. The implications of this discovery for how we think about cells' capabilities, and for the future treatment of degenerative and other diseases, are potentially profound." In other words, **when cells grow up, they remember their childhoods.** This represents one very practical future application of my cellular memory proposition.

Stem cells are undifferentiated cells that can turn into specific cells, as the body needs them. Stem cells originate from two main sources: adult body tissues and embryos. Most tissues have small reservoirs of stem cells that can replenish cells as they age or die. Unlike other cells in the body stem cells can become almost any cell that is required. Thus, they help our tissues recover from injuries.

Stem cells have traditionally been considered developmental blank slates. Recent work suggests instead that they may store memories of past wounds or inflammations so that they can promote more effective healing in similar situations in the future. New studies in the skin, gut, and airways suggest that stem cells, often in partnership with the immune system, **can use these memories to improve the responses of tissues to later injuries** and pathogenic assaults. For example, inflamed skin on mice that was allowed to heal was found to heal 2.5 times faster the next time at that same spot.

Shruti Naik, an immunologist at New York University who has studied this memory effect in skin and other tissues, has commented on the stem cells' excellent ability to sense their environment and respond to it. Stem cells also appear to communicate with the immune system to work as a team. And most important, these stem cells transfer their memories to future generations of cells.

I thought this is as far back as we can go in tracing the history of cells in our life. But a new study from Rockefeller University has proven me wrong. It does that by taking us back to conception when we all start out as a cluster of identical cells. As these cells divide and multiply, they gradually differentiate and become precursors of muscle cells, bone cells, neurons, etc. Previously, it has been shown that a network of signaling proteins—including one called *WNT*—trigger human embryonic stem cells to form the initial clusters. Then, another signaling protein called *ACTIVIN* tells the cells to specialize to form the two inner germ layers.

In experiments that applied only the ACTIVIN signal, the clump of embryonic cells kept dividing, making more stem cells instead of differentiating. The scientists repeated the experiments on cells that had previously been exposed to WNT signaling proteins. The cells behaved in the usual manner, specializing and forming the inner two germ layers. This suggests that the cells somehow remembered the WNT signal, and this memory changed how they responded to ACTIVIN. The next step is to understand how cells store the memory of the WNT signal.

As we have already seen when studying the immune system, it is becoming clear that neural networks have no monopoly on such functions as "subtraction, addition, gain control, saturation, amplification, multiplication, and thresholding." **Single somatic cells are able to perform the same or very similar functions. Furthermore, neural-like computation, decision-making, and memory have been observed in a wide range of**

systems well beyond the traditional CNS, including sperm, amoebae, yeast, plants, bone and heart, subjects we shall learn about in the following chapters.

All of the major mechanisms by which nerves function—ion channels, neurotransmitters, and electrical signaling exist in somatic cells. Exciting work on bioelectricity in somatic cells has led Michael (Mike) Levin, professor of biology at Tufts University, to suggest that **memory might be distributed throughout the body** with somatic cells communicating bioelectrically with each other through gap junctions (synapses) thus forming a network similar to neuronal networks capable of encoding information and directing cell activity.

Going a step further, for a long time it was believed that the brain governs all aspects of sleep. However, a new study by Joseph S. Takahashi, chairman of neuroscience at the University of Texas Southwestern Medical Center proves this assumption wrong. His research focused on a circadian clock protein found in mice muscles—*BMAL1*. Mice with higher levels of BMAL1 in their muscles were found to recover from sleep deprivation more quickly than a control group of mice. Equally telling, removing BMAL1 from muscle tissue severely disrupted normal sleep, leading to a reduced ability to recover. "This finding is completely unexpected and changes the ways we think sleep is controlled," said Takahashi. By demonstrating that factors in muscles can reach the brain and so influence sleep this research contributes one more robust proof of the interaction of body cells and the brain. In other words**, the brain is not the only game in town.**

DNA, Our Personal iCloud

Genetic research shows that chemical reactions in the primordial soup created increasingly complex RNA molecules from which viruses evolved. Viruses were not only the probable precursors of the first cells, but a major factor in evolution. As we peer into the rearview mirror of humanity's earliest beginnings, we can apply this same technique to matters closer to our own lives. I am thinking here of the rise in direct-to-consumer DNA testing.

Consider this story that appeared about AncestryDNA, a commercial company that will analyze your DNA for a fee. One day, a man named Barry logged into Ancestry and found a parent-child match with a young woman. He was floored by the news that he had a daughter. Eventually he arranged a Thanksgiving dinner with her, and met her. Apart from being very happy about their reunion, they learned of one more surprise: they both turned out to be genealogists.

DNA is a gateway to the long line of ancestors who made our existence today possible. I believe with my whole heart that the more we know about ourselves and the more we understand the many elements that have affected our lives, the better we can live a life of reason, joy, and caring. Though there may be surprises along the way, most people will find the journey rewarding.

DNA as we have already learned resides in the cell nucleus. For a long time, the nucleus of the cell, which is enclosed in a membrane, was considered to be the "command center" of the cell and was deemed the cellular equivalent of the "brain." However, this hypothesis is now being challenged by a number of scientists who believe that it is the cell membrane of the entire cell that represents the brain of the cell and not the nucleus.

According to cellular biologist Bruce Lipton, one of the leading members of this brave new breed of scientists, the cell membrane contains receptor and effector proteins. Receptors function as molecular nano-antennas (*nano* refers to very small items) that monitor both internal and external states. The membrane's receptors function like sensory nerves, and the effector proteins like action-generating motor nerves. It is therefore quite reasonable to compare, as Lipton and others have, the cell's membrane *with its gates and channels* to an information-processing *transistor* and organic *computer chip* or *microchip*. Therefore, the membrane of all cells may be another place where memories/data may be stored.

In the past thirty years, the *nanomicroscope* (atomic force microscope, or AFM) has revealed how proteins extract energy from the environment and perform the tasks needed to keep an organism functioning. It is with the aid of these and more advanced microscopes that the science of *nanotechnology* was born. Nanotechnology explores the nanoscale world of individual molecules. Nanotechnologists work on producing hybrid bioinorganic devices that mimic biological processes destined to be used in new computers and electronic devices, particularly microchips.

A microchip, as most people know, is a set of electronic circuits on one small plate ("chip") of semiconductor material, normally silicon. Integrated circuits are used in virtually all electronic equipment today and have revolutionized the world of electronics. Computers, mobile phones, and other digital home appliances are now inextricable parts of the structure of modern societies, made possible by the low cost of producing integrated circuits. Microchips can be tightly packed with several billion transistors and other electronic components in an area the size of a fingernail. The width of each conducting line in a circuit will grow smaller and smaller as the technology advances.

In 1997 an Australian research consortium headed by B. A. Cornell from the Australian National University, Canberra, published an article in *Nature* that showed that the cell membrane not only looks like a computer chip but actually acts like one. Cornell and associates successfully turned a biological cell membrane into a digital readout computer chip.

Another Australian study, this time from the Functional Materials and Microsystems Research Group (RMIT) in Melbourne, has succeeded in constructing the world's first electronic long-term memory cell using a functional oxide material in the form of an ultra-thin film—ten thousand times thinner than a human hair. Project leader Sharath Sriram is quoted as saying, "The ability to create highly dense and ultra-fast analog memory cells paves the way for imitating highly sophisticated biological neural networks." In other words, biological neural networks are still smaller and faster than any device the world's best scientists can construct today.

In 2010 Taiwanese scientists unveiled a new computer microchip, which was at that time the smallest widget of its kind ever manufactured, measuring just nine nanometers across (one nanometer is equal to just one billionth of a meter). Laboratory director Yang Fu-liang said that using this technology, a one-square-centimeter chip would be able to store one million pictures or one hundred hours of 3D movies.

In 2015 a team from the University of Michigan built not just a very small microchip but also the world's smallest functioning computer, less than a cubic millimeter in size. Nearly 150 of these computers fit inside a single thimble. Called the Michigan Micro Mote, or M3, this nano computer features processing, data storage, and wireless communication. The chips are designed to work with other chips, collectively termed *smart dust*.

Toward the end of 2020, engineers at the University of Texas created the smallest memory device yet. They shrank the cross section of a microchip area down to just a single square nanometer. The scientific holy grail for scaling is tumbling toward a level where a single atom now controls the memory function. By creating ever smaller chips engineers aim to decrease these chips' energy demands and increase their capacity, which means faster, smarter devices that take less power to operate.

How is this relevant? Because these discoveries show the incredible amount of information that can be packed into a tiny space, and by analogy, allows for the same process to operate in living cells.

At the Massachusetts Institute of Technology (MIT), Timothy Lu, a member of the Synthetic Biology Group, and others are engineering circuits into bacterial cells, literally programming them for data storage and computation. And since each position of DNA can encode four different pieces of information—cytosine (C), guanine (G), adenine (A), and thymine (T)—instead of just two, as with classic binary silicon systems, DNA could someday, in principle, store more data in less space. The same properties that make DNA a great genetic code for living organisms also makes it a desirable substrate for data storage on computers.

In 2013 Nick Goldman and his group from the European Bioinformatics Institute already managed to encode all of Shakespeare's sonnets into DNA. As if that wasn't enough, in 2017 researchers from Harvard Medical School galvanized the scientific world by encoding into the genomes of living bacteria five frames from a classic 1870s "race horse in motion" sequence of photos—an early forerunner of movies (FIG 4.3). "This work demonstrates that this system can capture and stably store practical amounts of real data within the genomes of populations of living cells," wrote the lead author of the study, Seth L. Shipman.

George Church, a geneticist at Harvard University and one of the authors of the new study, recently encoded his own book, *Regenesis*, into bacterial DNA and made ninety billion copies of it. "A record for publication," he said wryly in an interview. **DNA, as a compact and stable information-storage medium is a transformative new concept in biology and computer technology.**

Obviously, DNA has the potential to store massive amounts of information. At the present, we know what some of this information is, but much of it remains terra incognita. It is like the dark side of the moon. It may not be visible, but we know it exists. I hypothesize that memories about our life and our ancestors' lives are stored in this unexplored area of DNA, the cellular membrane and the cytoskeleton.

Summary

The human cell is an amazing biological entity. It is hard to truly appreciate how something so small can contain so much and serve so many functions. It reminds me of a *matryoshka* doll, also known as a Russian nesting doll (FIG 4.4), a set of wooden dolls of decreasing size placed one inside another. Imagine that the outermost doll represents the whole universe, inside of which is inserted our galaxy, then the earth, then humans, then cells, then atoms, then subatomic particles. But, unlike the Russian nesting dolls, **all these systems, whether we are conscious of it or not, are connected and affect each otherall the time.**

We live in an age where most of us understand microchips and computers better than the biology of cells. Over the last half century, we watched how computers that occupied enormous spaces in university labs and functioned with very little memory shrunk in size—some now as small as 1/16 mm—while increasing their memory exponentially. We now see microchips storing enormous amounts of memory in a very small space.

The cell membrane, which possesses all the characteristics of microchips, in addition to performing other duties, also directs the production of proteins that encrypt memories. This would apply to all cells in the body; in other words, to both cortical cells and somatic cells (body cells). While the entrenched scientific view holds that memories in the brain are located in the synapses, evidence in the chapter on the brain has shown convincingly that engrams are stored in neuronal cells.

Today, **intelligence** is generally defined as the ability to understand and adapt to the environment by using inherited abilities and learned knowledge. In order to apply learned knowledge, it is necessary to have memory. We have seen that this is in fact the case. There are at least three places in a cell where memories are located: the cell membrane, the cytoskeleton with its nanotubules, and the cell nucleus with its DNA and RNA. This is the case with all cells including stem cells.

Cells communicate with neighboring cells and provide them with aid in case they suffer from stress or sickness. Adult tissues retain a memory, inscribed on their DNA, of the embryonic cells from which they arose. Let us not forget about the axons and dendrites of neurons from the chapter on the brain. We learned that the growth cones of axons contained much of the molecular machinery of an independent cell, including proteins involved in growth, metabolism, signaling, and more, and that tiny compartments in the dendritic arms of cortical neurons can each perform complicated operations in mathematical logic.

Based on the evidence, as counterintuitive as this may be, we can conclude that the cells in our bodies are truly intelligent and as such, form an essential and necessary substrate of the embodied mind.

KEY TAKEAWAYS

- The cell membranes of cortical and somatic cells store information/memory.
- Information/memory is also stored within the cell.
- DNA is the cellular equivalent of the iCloud.
- Each cell has enormous capacity for information storage.
- All of the major mechanisms by which nerves function—ion channels, neurotransmitters, and electrical synapses—exist throughout the body.
- Biological cell assemblies work in concert with neural networks.
- Cells remember their origins, all the way back to conception.

CHAPTER FIVE

THE INTELLIGENCE OF SINGLE-CELLED ORGANISMS

Introduction

Unicellular organisms are believed to be the oldest form of life, possibly going back 3.8 billion years. Studying unicellular organisms is treacherous territory for a scientist. It is a minefield of unexploded, unanswered questions. One of the difficulties is to define life, or to differentiate between plants, organisms, and animals.[1] Experts disagree about whether viruses are "alive," or, much more controversially, whether prions are alive. However, there is widespread agreement that bacteria are the smallest, indisputably living things on earth.

We have spoken of what constitutes intelligence in the previous chapter. On the basis of that definition, scientists are now starting to recognize intelligence in various unicellular and multicellular animals. These organisms or animals, because they are that, have no brains, not even neurons. Yet, they seem capable of learning, decision-making, goal-directed behavior, and memory, at least when it comes to viruses that attack them. They sense and explore their surroundings, communicate with their neighbors, and adaptively reshape themselves.

Great advances in molecular medicine in recent years have led to the discovery of an immense number of microorganisms in the intestine referred to as the *gut microbiome.* In parallel with learning about bacteria in general and the bacteria in the gut specifically, we shall also explore the concept of the brain-gut axis that links the central and the *enteric nervous system,* the nervous system of the alimentary canal. Along this route exchange of information takes place in both directions.

Bacteria

Typically, bacteria are shaped like spheres, rods (FIG 5.1), or spirals. *Escherichia coli*, the bacteria responsible for fecal contamination in food is about 7 μm (7 microns) long and 1.8 μm in diameter. On the average there are 40 million bacterial cells in a gram of soil and one million bacterial cells in a milliliter of fresh water. There are approximately 5×1030 bacteria on earth, forming a *biomass* that exceeds that of all plants and animals combined.

Once described as mere "bags of enzymes," bacterial cells were long thought to have little to no internal structure. Bacteria, as opposed to single-celled organisms like the slime mold that we shall discuss shortly, do not have a membrane-bound nucleus, and their genetic material is typically a single circular DNA chromosome with its associated proteins and RNA located in the cytoplasm in an irregularly shaped body called the *nucleoid* (FIG 5.2). Very recently, researchers at McGill University have discovered bacterial organelles involved in gene expression, suggesting that bacteria may not be as simple as once thought. In *Eschericha coli*, the bacteria that were the subjects of the McGill study, organelles are held together by "sticky" proteins rather than a membrane. Scientists refer to this system as *phase separation*. It may be a universal process in all cell types, and it may have been involved in the very origin of life on earth.

Bacteria face a never-ending battle with viruses and invading strands of nucleic acid known as *plasmid*s. To survive this onslaught, bacteria and *archaea*[2] deploy a variety of defense mechanisms, including an adaptive-type immune system that revolves around a unit of DNA known as *CRISPR*, which stands for *Clustered Regularly Interspaced Short Palindromic Repeats.*

The bacterial immune system also relies on an enzyme called *Cas9* that uses immunological memories to guide cuts to viral genetic code. It seems that *Cas9* directs the formation of these memories among certain bacteria.

Researchers at Berkeley's Department of Molecular and Cell Biology have observed for the first time that through the combination of CRISPR and squads of CRISPR-associated—"Cas"—proteins, bacteria are able to utilize small customized RNA molecules to silence critical portions of a foreign invader's genetic message and acquire immunity from similar invasions in the future by "remembering" prior infections. It works by altering the bacterium's genome, adding short viral sequences called spacers in between the repeating DNA sequences. **These spacers form the memories of past invaders**. They serve as guides for enzymes encoded by CRISPR-associated genes (Cas), which seek out and destroy those same viruses should they attempt to infect the bacterium again.

Other researchers at Berkeley studied a strain of *Bacillus subtilis*'s capacity to "remember" ten distinct cell histories prior to submitting them to a common stressor. The analysis—much too complicated to describe here—suggested that Bacillus subtilis remembers, for a relatively long time, aspects of its cell history. Though unable to explain the underlying biological mechanisms, the scientists have generated an information-theory-based conceptual framework for measuring both **the persistence of memory in microbes and the amount of information about the past encoded in their organisms.**

Take the extremophile *Deinococcus radiodurans*, an organism that thrives in physically or geochemically extreme conditions, detrimental to most life on earth, and drop it in acid and it survives. Eject it into outer space and it thrives. Desiccate it; store it for a million years then pull it out, and the bacterium will come back to life. *D. radiodurans* suffers few—if any—mutations, even when blasted with ionizing radiation. It's been nicknamed Conan the Bacterium by NASA. Because this bacterium will survive any pandemic, global warming or cooling, or atomic warfare, computer scientist Pak Chung Wong at the Pacific Northwest National Laboratory inserted the lyrics of the song "It's a Small World After All" encoded in DNA, into the genome of this bacterium. Then he allowed it to multiply for one hundred generations. The embedded song survived without mutations. In other words, memory survived one hundred generations. Here we see bacteria serving as information time capsules.

Bacteria in our bodies have a choice: they can stay single and separate from their families and friends and expose themselves to attack by their host's immune system, and perhaps by antibiotics, or band together with their fellow travelers and form a *biofilm,* thus greatly improving their chances of survival. Biofilms are three-dimensional structures made up of polysaccharides secreted by the bacteria, recycled DNA, and materials from dead or dying bacteria. But biofilms are not just dense accumulations of bacterial cells. They have complex functional structures, inside and out, that serve the cells' collective life.

Over time the bacterial cells in specific regions of the biofilm will start to interact with their neighbors in different ways. Bacteria communicate through a chemical process called *quorum sensing,* in which they release molecules that serve as messages detected by nearby bacteria by way of the previously discussed nanotubes. These nanotubes have also been detected to operate in an interspecies manner, between *B. subtilis*, *Staphylococcus aureus*, and *Escherichia coli*.

Quorum sensing can occur within a single bacterial species as well as between diverse species. Bacteria use quorum sensing to coordinate certain behaviors such as virulence, antibiotic resistance, and biofilm formation.

Gürol M. Süel and his team at the University of California San Diego have studied long-range communication in biofilms for many years. They have reported observing waves of charged ions propagated through the biofilm to coordinate the metabolic activity of bacteria in the inner and outer regions of the biofilm. "**Just like the neurons in our brain . . . bacteria use** *ion channels* **to communicate with each other through electrical signals,**" said Süel.

Bacteria in biofilms are responsible for many chronic diseases. Naturally, they are very resistant to all forms of drugs. Scientists from MIT have for the first time observed two strains of bacteria that are each resistant to one antibiotic protecting each other in an environment containing both drugs. The findings deepen our understanding of *mutualism*, a phenomenon more commonly seen in larger animals, in which different species benefit from their interactions with each other. This cross-protection can help bacteria form drug-resistant communities.

According to a collaborative group of chemists, biologists, physicists, and engineers working together at Australia's Flinders Biofilm Research & Innovation Consortium, bacteria are equipped with sensory systems that connect with cognitive-behavioral circuits and show many other neural features.

Since 1983, Roberto Kolter, a professor of microbiology and immunobiology at Harvard Medical School, has led a laboratory that has studied these phenomena. Kolter has stated that under the microscope, **the incredible collective intelligence of bacteria** reveals itself with spectacular beauty. Imagery from the lab, World in a Drop, has been on display in an exhibition at the Harvard Museum of Natural History.

In view of the above and other related studies many leading cognitive scientists have now concluded that **communities of bacteria within the biofilm appear to act as a kind of** ***microbial brain*****.**

One other point about bacteria. You may recall a discussion we had about mitochondria in the chapter on the cell. This organelle is the ancient remnant of symbiotic bacteria that invaded host cells about two billion years ago and became specialized in energy production. Mitochondria still carry a small amount of DNA of their own, although with just thirty-seven genes, they have less genetic material than any living bacteria. Like bacteria these closely packed mitochondria change in size and shape and some of them also form linkages, similarly to bacteria, for sharing information. Recently, several studies have revealed that mitochondria might be essential not just to our general physical well-being but specifically may play a surprisingly pivotal role in mediating anxiety and depression. This is another example of how organelles in nonneuronal body cells affect our thinking and feeling.

The Microbiome

The human gastrointestinal tract (GIT) is populated with as many as one hundred trillion bacterial cells. Their collective weight may exceed five pounds, about 2.2 kilograms. Recently there has been a groundswell of research into how these gut bacteria influence critical aspects of our physiology. We are learning that the myriad resident gut microbes are critical for the development and function of the immune system. Even more important, gut microbes affect the complex and distant brain. This seemingly improbable concept has been well established in the past decade. Findings from animal studies have shown that the *microbiota* (the sum of all the bacteria, fungi, and viruses inhabiting our gut) can affect brain metabolites, behavior, and neurogenesis, that is, the formation of new neurons.

Gut microbes contain 3.3 million genes, dwarfing the human genome's twenty-three thousand genes. These microbiota genes produce a myriad *neuroactive compounds*. Gut bacteria contribute to the host metabolism by production of metabolites such as bile acids, choline, and short-chain fatty acids (SCFAs) that are essential for host health. SCFAs are a subset of fatty acids that are produced by the gut microbiota during the fermentation of complex carbohydrates such as dietary fiber and partially digestible and nondigestible polysaccharides. The highest levels of SCFAs are found in the proximal colon, where they are locally absorbed and transported by the gut epithelium (inner intestinal cells) into the bloodstream and eventually to the brain.

Signals from the gut travel to the rest of the body along the hypothalamic-pituitary-adrenal axis (HPA). However, the vagus nerve is the primary route that enables stimuli from the brain to pass to the gut and from the gut to the brain. Only recently has it been established that the vagal gut-to-brain axis is also an integral component of the neuronal reward pathway. Researchers from Yale University, Duke University, and the University of São Paulo working jointly discovered for the first time, that the right and left branches of the vagus nerve ascend asymmetrically into the central nervous system. Only the right branch of the gastrointestinal vagal nerve conveys reward signals to the dopamine-containing reward neurons in the brainstem.

In addition to the vagus nerve, the brain-gut-microbiota axis (FIG 5.3) includes the central nervous system, the sympathetic and the parasympathetic system, the neuroendocrine and immune system, and the enteric nervous system.

The enteric nervous system (ENS) is the intrinsic nervous system of the gut. It consists of an extensive network of neurons that lines the walls of the gastrointestinal tract. The complexity of the human ENS is only exceeded by the brain and spinal cord.

The ENS has the unique ability to control the behavior of its organ, the gut, without input from the central nervous system. It was this independence of the ENS that led Michael Gershon at Columbia University to call it "The Second Brain." Biologist Michael Schemann, with colleagues at several German universities, documented the presence in the ENS of habituation, sensitization, conditioned behavior, and long-term facilitation. As such, the ENS is capable of various forms of implicit learning and remembering. In his view, the ENS represents a "smart" system.

It has been shown that a faulty composition of the gut microbiota in childhood influences the maturation of the central nervous system. Numerous studies indicate that the dysfunction of the brain-gut axis can lead to both inflammatory and functional diseases of the gastrointestinal tract. It is becoming increasingly clear that psychiatric and neurological illnesses, including multiple sclerosis, autism, schizophrenia, and depression, are often present simultaneously with gastrointestinal disease.

One of the most common gastrointestinal diseases is irritable bowel syndrome (IBS). An intriguing feature of IBS is its frequent comorbidity with other disorders and symptoms. Research in Norway on patients with irritable bowel has shown that while the prevalence of mood disorder is 11 percent in the Norwegian population, it is 38 percent in Norwegians diagnosed with IBS.

The gastrointestinal tract is now also recognized as a major regulator of motivational and emotional states. Gastrointestinal problems are common among people with depression and anxiety, and studies suggest people with depression have a different gut flora from emotionally healthy people.

Intestinal bacteria also produce serotonin, dopamine, and other brain chemicals that regulate mood. The hope is that enhancing good gut microbes—whether with probiotics, fecal transplants, or capsules filled with donor stool, or by adding sauerkraut or other fermented foods to the diet—may be the answer to intractable depression, the kind conventional treatments can't touch. It could also fundamentally alter the way we conceptualize mental illness. For the longest time medical experts believed that mental illness was essentially a brain illness, when in fact it may be much more complex than that.

Therefore, it is not surprising that the scientific literature is increasingly reporting research on the effects of psychotropic drugs on intestinal microbes. Recently, a group of Ireland-based scientists found that some drugs consistently increased the number of certain bacteria in the gut. For example, lithium and valproate (both used for bipolar disorder) increased the numbers of clostridium and other bacteria. Not a desirable outcome. On the other hand, the (SSRI) antidepressants escitalopram and fluoxetine significantly inhibited growth of bacterial strains such as *E. coli*. This is good and benficial.

It is likely that psychotropic drugs, in fact all drugs, may affect intestinal microbes as part of their mechanisms of action. Depending on a person's microbiome, certain drugs may benefit some people but not others. No doubt, assessing an individual's microbiome before commencing treatment will be an important lab test in the future.

New research shows another connection between the gut and the brain. A diet rich in salt has been recognized to increase the risk of cerebrovascular diseases and dementia. Up to the present it was unclear how dietary salt harms the brain. Now, scientists at Weill Cornell Medicine in New York unveiled a previously unknown linkage between increased salt intake and cognitive impairment through a gut-initiated adaptive immune response. Dietary salt seems to promote neurovascular and cognitive dysfunction through a gut-initiated *T-helper cell 17* (Th17), a recently discovered subset of effector memory T cells. Th17 cells are hallmarks of many immune-inflammatory diseases.

The great majority of human gut microbes are friendly rascals that make important contributions to our health. Bioengineers at Baylor University have recently begun to explore the influence of gut bacteria on the aging process. In experiments on our old friend, C. elegans, they induced *E. coli* in the worm to produce more *colanic acid*, which protected the worm's gut cells against stress-induced mitochondrial fragmentation. Mitochondria have been increasingly recognized as important players in the aging process. They observed that worms carrying this *E. coli* strain lived longer.

We are learning that for optimum brain development of an infant to occur, the initial colonization of their gut must be derived from their mother's birth canal. Cesarean delivery is a largely unrecognized threat to the microbial handoff from mother to child. Instead of traveling down the birth canal picking up lactobacilli, the baby is surgically extracted from the womb. Infants delivered by caesarean section are colonized by bacteria dominated by *staphylococcus*, *corynebacterium*, and *propionibacterium* from environmental sources, including health-care workers, air, medical equipment, and other newborns. These bacteria could adversely affect their brain development and their ability to digest milk.

Another way in which gut bacteria have a positive effect on unborn children is by the production of multiple *low-molecular weight (LMW)* substances in their blood. These microbial LMW substances may be one of the key internal factors regulating optimal human genetic expression throughout life.

However, the beneficial impact of the pregnant mother's microbiome[3] will be undercut if she frequently drinks from plastic bottles or eats from cans containing bisphenol A (BPA). New evidence from a research study in rabbits suggests that BPA exposure just before or after birth leads in the newborn to reduced gut bacterial diversity, a decrease in bacterial metabolites, such as short-chain fatty acids (SCFAs), and elevated gut permeability—three common early markers of inflammation-promoted chronic diseases. Impaired microbiota may foster chronic colon and liver inflammation.

Studying bacteria, we find credible evidence for host-microbe interaction at virtually all levels of complexity, ranging from direct cell-to-cell communication to extensive systemic signaling, and involving various organs and organ systems, including the central nervous system. As such, the discovery that differential microbial composition is associated with changes in thinking, feeling, and behaving has significantly contributed to extending the well-accepted gut-brain axis concept to the microbiota-gut-brain axis.

The microbiome and its relationship to the enteric nervous system, the brain, and the rest of the body is a perfect example of one of the many unrecognized parts of the Embodied Mind. Out of sight is not out of mind.

Protoplasmic Slime Molds

Slime molds arrived on land close to a billion years ago. They may well have colonized continents that were home only to films of bacteria. Slime molds live in soil. They may spend their lives as single-celled organisms or grow into multicellular, gigantic, pulsating networks of protoplasm (FIG 5.4). John Tyler Bonner, emeritus professor in the Department of Ecology and Evolutionary Biology at Princeton University, is one of the world's leading experts on cellular slime molds. Developmental-evolutionary biology owes a great debt to the work of Bonner. I had the good fortune of corresponding with Professor Bonner, who passed away at the age of eighty-seven in 2019. His work is clearly written and easily accessible to an informed reader. I highly recommend it to anyone wishing to peruse this subject further.[4]

In his 2009 book, *The Social Amoebae*, Bonner described slime molds as very different from other organisms; they feed as individual amoebae before coming together to form a multicellular organism that has a remarkable ability to move and orient itself in its environment. Furthermore, these social amoebae display a sophisticated division of labor; within each organism, some cells form the stalk and others become the spores that will seed the next generation.

The favorite of scientists who study slime molds is the Physarum polycephalum, a large amoeba-like cell with multiple nuclei and a dendritic network of tubelike structures (pseudopodia). It should be emphasized that like all molds, this one is brainless. In fact, it does not even possess neurons. Slime molds may range from only a few millimeters in diameter to well over 12 inches (30 cm) across. The plasmodium moves like a giant amoeba, flowing over the surface as it ingests dead leaves, wood, bacteria, and microbes. It also secretes enzymes for digesting the engulfed material.

"Physarum polycephalum" follows fairly basic behavioral patterns, growing toward food and away from light. Biologists at Macquarie University in Sidney demonstrated that Physarum polycephalum constructs a form of spatial memory by avoiding areas it has previously explored. This mechanism allows the slime mold to build an effective and robust food transportation network. Their research **provides a unique demonstration of a spatial memory system in a nonneuronal organism**.

Physarum polycephalum changes its shape as it crawls over a plain agar gel and, if food is placed at two different points, it will put out pseudopodia[5] that connect the two food sources. This simple organism has the ability to find the minimum-length solution between two points in a labyrinth. It can solve mazes and can closely approximate human transport networks on flat and three-dimensional surfaces. Does that imply analytic thinking, as we know it? I think the answer has to be in the affirmative.

Researchers at the Tokyo Institute of Technology conducted a series of control experiments by dividing one individual Physarum polycephalum mold that had been trained into two separate organisms and had them perform parallel searches for food. This experimental design reminds me of similar experiments on the planarian roundworm we shall consider in the next chapter. The two molds from the original trained one found a solution faster than two similar untrained molds. In their published paper the scientists wrote that their results **signify the presence in this organism of long-term memory as well as an ability to perform sophisticated computing tasks.**

Using a different experimental design when utilizing multiple separate food sources, researchers at Japan's Hokkaido University concluded that **the plasmodium tube network is a well-designed and intelligent system.** Not surprisingly, one of the leading scientists in this field, Soichiro Tsuda of Kobe University, has suggested that Physarum polycephalum demonstrates **emergent intelligence.**

In other experiments at Hokkaido University, scientists subjected Physarum polycephalum to a series of shocks at regular intervals. It quickly learned the pattern and changed its behavior in anticipation of the next shock. This memory stays in the slime mold for hours, even when the shocks themselves stop. A single renewed shock after a "silent" period will leave the mold expecting another to follow in the rhythm it learned previously. T. Saigusa and his colleagues say their recent findings **"hint at the cellular origins of primitive intelligence."** Microbiologist James Shapiro of the University of Chicago, commenting on their paper said that if the results stand up, **"this paper would add a cellular memory to those capabilities."**

The Good News

Many physiological and pathological features of pregnancy are controlled, at least in part, by the mother's resident microbes, which evolved to help her and themselves. When food is in short supply during pregnancy, as has often happened in human history, the mother's microbes will shift their net metabolism so that more calories flow from food to her body.

Also, women of reproductive age carry bacteria, primarily lactobacilli, which make the vaginal canal more acidic. This environment provides a hardy defense against dangerous bacteria that are sensitive to acid. The first fluids a baby sucks in contains their mother's microbes, including lactobacilli. Lactobacilli and other lactic acid–producing bacteria break down lactose, the major sugar in milk, to make energy. This is good news for vaginally born children.

Gut microbes produce compounds that prime immune cells to destroy harmful viruses in the brain and nervous system. Notably, a study from the Department of Pathology at University of Utah Health, Salt Lake City, has shown that mice treated with antibiotics (which kill the "good" bacteria) before the onset of disease were unable to defend themselves. These mice had a weak immune response, were unable to eliminate the virus, and developed worsening paralysis, while those with normal gut bacteria were better able to fight off the virus.

It follows that having a healthy and diverse microbiota is crucial for quickly clearing viruses from the nervous system to prevent paralysis and other risks associated with diseases such as multiple sclerosis. Therefore, if you have been prescribed antibiotics, it is imperative that you restore your microbiome to its former healthy state by either consuming foods rich in friendly bacteria, such as yogurt, kefir, or sauerkraut, or taking probiotics.

Two studies—one in mice and the other in human subjects—point to the importance of exercise in keeping a healthy gut microbiome. Jeffrey Woods, a University of Illinois professor of kinesiology, showed that exercised mice had a higher proportion of microbes that produce *butyrate,* a short-chain fatty acid that promotes healthy intestinal cells, reduces inflammation, and generates energy for the host. They also appeared to be more resistant to experimental ulcerative colitis, an inflammatory bowel disease. This effect was more pronounced in lean compared to obese adult mice.

We have known for some time that avocados help you feel full and reduce blood cholesterol. But now scientists at the University of Illinois found that people who ate avocados every day as part of a meal had a greater abundance of gut microbes that break down fiber and produce metabolites that support gut health. They also had greater microbial diversity compared to people in the control group. Avocados are the new apples. Be the first among your friends to say, "An avocado a day keeps the doctor away."

Also, maintaining an optimal vitamin D concentration appears to be beneficial for gut health. Vitamin D modulates intestinal microbiome function, controls antimicrobial peptide expression, and has a protective effect on epithelial barriers in the gut mucosa. So, make sure you receive sufficient vitamin D through sunlight or supplementation.

Summary

Studies of cognition in animals have been routinely impeded by anthropocentrism; that is, by the assumption that human thinking is the benchmark by which animal cognition should be measured. Such an approach tends to ignore the incredible diversity of intellectual abilities present in living organisms. A realignment of focus demonstrates the presence of faculties of memory retention, navigation, communication, pattern recognition, and statistical reasoning in brainless unicellular organisms such as bacteria and slime molds among other ancient organisms.

Bacteria—often viewed as lowly, solitary creatures—are actually quite sophisticated in their social interactions and communicate with one another through electrical signaling mechanisms similar to neurons in the brain. When these unicellular organisms commune in great numbers, their remarkable collective talents for solving problems and controlling their environment emerge even more sharply.

There are two important conclusions we can draw from the research cited here on bacteria and slime molds. First, unicellular organisms have been shown capable of interpreting the world as either nutritious or toxic. Tiny bacteria or slime molds, brainless and neuronless, demonstrate evidence of **rudimentary intelligence**. Surely the cells in our bodies that are larger and more complex share or even exceed this intelligence.

The second conclusion is that the bacteria in the gut affect our brain, the ENS, and the rest of our bodies in a variety of important ways, until recently largely unrecognized. **This microbiome-brain axis and the ENS–immune system linkage represent two of the essential constituent parts of the Embodied Mind.**

Taking into account the crucial part that the "Thoughtful Bowel" plays in our behavior and thinking, when somebody says, "I had this gut feeling," don't laugh it off and reflexively dismiss it.

KEY TAKEAWAYS

- **Bacteria are capable of learning and remembering.**
- **Bacterial communities (biofilms) process information and make decisions about nutrient distribution and metabolism as an integrated whole, using ion channels similar to the way neurons operate in the brain.**
- **The vagal gut-to-brain axis plays a critical role in motivation and reward.**
- **The microbiome has multiple critical effects on our physiological and metabolic processes ranging from prenatal brain development and modulation of the immune system to, perhaps most surprisingly, behavior and cognition.**
- **A healthy gut microbiota contributes to normal brain function.**

- The slime mold Physarum polycephalum hints at the cellular origins of primitive intelligence.

- There is no mistaking the fact that processes and activities thought to represent evolutionary "recent" specializations of the nervous system represent ancient and fundamental cell survival processes that persist in all our cells.

CHAPTER SIX

REGENERATION, HIBERNATION, AND METAMORPHOSIS

Introduction

It seems like from the start of civilization it has been assumed that one of the fundamental functions of the brain is its ability to store memories, thus allowing animals, including humans, to alter behavior in light of past experience. If the seat of all memory is in the brain, to ensure long-term stability of encoded information the brain cells and their circuits must remain stable like the books on your bookshelf. If someone started to tear pages out from these books, not only would they be seriously damaged but you would forever lose your books' contents. In view of this, animals such as the planaria that exhibit a remarkable capacity to quickly regrow new body parts along with their brains, confront us with a fascinating question: How can fixed memories persist in the planaria when their brains are removed and new brains grow out of their previous body parts?

Similarly, animals that hibernate and undergo massive pruning of their cerebral neurons during the cold months confront us with a similar problem.

Because when they recover their strength and health in the spring, many of their previously learned behaviors return.

Metamorphosing animals[1] like the common frog develop from a larva into a tadpole and then into a frog. Caterpillars go through five stages of growth. Yet, as we shall see, memories formed in these animals in their earliest embryonic states survive extensive remodeling of their bodies, including their brains.

Regeneration

Planarian flatworms, on account of their regenerative capabilities, are one of nature's gifts to science. Splitting a planarian down the middle quickly gives rise to two cross-eyed little worms staring back at you. Decapitation leads to the development of two new worms. Whole worms can regenerate from only small slices of the adult worm within few days. Some biologists have succeeded in chopping up one planarian into more than 200 pieces. Each tiny piece eventually formed a miniature complete worm, which grew in time to its normal size of up to ¾ inches, depending on the species and the availability of food (FIG 6.1).

How do planaria manage such an incredible feat? A few years ago, researchers discovered that a resident population of adult stem cells (*neoblasts*) enable these worms to regenerate any body part after surgical removal of that part. However, as important as stem cells are, they cannot account for the persistence of memory after a planarian's head is removed and its body grows a new head.

The planarian is one of the simplest animals living on this earth with a body plan of bilateral symmetry and head orientation. The brain of these flatworms has a bilobed structure with a cortex of nerve cells and a core of nerve fibers including some that connect the two hemispheres. Special sensory signals from chemoreceptors, photoreceptor cells of primitive eyes, and tactile receptors are integrated to provide local reflexes and motor responses. Many structural features of planarian neurons, including synapses, are similar to human brains. Neurotransmitters identified in the planarian nervous system also occur in the human brain (FIG 6.2). Because of their extensive regenerative capacity (driven by an adult stem cell population) and complex CNS, significant efforts are underway to understand the molecular mechanisms behind neural repair and patterning. Planaria are thus a popular organism for the study of memory.

In the 1950s and 1960s, experimental psychologist James V. McConnell and colleagues at the University of Michigan conducted studies using planarians to explore memory processes. In one series of experiments, planarians were trained to respond to certain stimuli, light and electric shock. When their heads were cut off and their bodies regenerated a new head, many of the regenerated worms demonstrated by their responses that they remembered their training.

In another series of experiments, planarians conditioned to respond to light-shock association were ground up and fed to other planarians. These cannibal worms learned to respond to the stimulus faster than a control group. McConnell interpreted this as evidence that **memory in flatworms was not localized in the head but was distributed throughout the animal's body.**

Many in the scientific community did not trust these experiments, citing problems with the use of appropriate controls, observer bias, and other reasons. But in 2013 a group led by Tal Shomrat and Mike Levin published a paper that essentially supported McConnell's findings and, by some reports, opened up a whole new can of worms. (Ouch, I could not help it. However, I promise it is the only pun in this book.)

On the occasion of welcoming Levin to Tufts University in 2008, dean of Arts & Sciences Robert Sternberg said, "Michael Levin represents a unique hire for the School of Arts and Sciences. His interests cross many fields, including biology but also engineering, computer science, neuroscience, and medicine. He represents the kind of distinguished interdisciplinary scholar who characterizes the essence of what makes Tufts a truly special place."

Because Levin's recent research has focused on **how living systems learn and store information in cells and tissues outside the brain**, his work is of particular relevance to us. Shomrat and Levin's 2013 paradigm-shifting paper reported on their experiments on planaria of the species *Dugesia japonica*. The researchers took advantage of a quirk of planarian behavior: once they get used to a familiar location, they will settle in to feed more quickly than planarians that find themselves in a new environment. Also, the worms naturally avoid light.

The researchers had one group of planarians living in containers with a rough textured floor while the other group was housed in a smooth-floored petri dish. After a few days the worms were tested to see how readily they would eat liver in an illuminated quadrant on the bottom of rough-textured dish. Automated video tracking and subsequent computer analysis of the worms' movements showed that the group that had spent time in the rough-floored containers overcame aversion to the light significantly more quickly and spent more time feeding in the illuminated space than did the non-familiarized group.

Both groups of worms were then decapitated and housed in a smooth-floored environment while their heads regenerated. Two weeks later, the fully regenerated segments were again tested. Worms regenerated from the familiarized group were slightly but not significantly quicker to feed in the illuminated part of the container, demonstrating that they retained recognition of the link between this type of surface and a safe feeding environment.

However, the worms exhibited no learned behavior prior to the regrowth of their brains. Evidently, the planarian needs to possess a brain for the behavior to occur. Takeshi Inoue, at Japan's Okayama University, hypothesizes that the new brain is regenerated as a blank slate and is gradually imprinted by traces of the previous memory by the worms' peripheral nervous system (which would have been modified during the training phase). Due to the modified peripheral nervous system, retraining the new brain during a short "saving" session suffices to restore full memory.

Shomrat and Levin, in their memorable 2013 paper suggested that **traces of memory of the learned behavior are retained outside the brain**. But rather than assuming that this is accomplished by the peripheral nervous system, they believe that it is by way of mechanisms that include the cytoskeleton, metabolic signaling circuits, and the gene regulatory networks. All of these exhibit (physiological) experience-dependent rewiring and rich feedback loops that can store information. They have indicated that **all of the major mechanisms by which nerves function—ion channels, neurotransmitters, and electrical synapses—exist throughout the cells and tissues of the body that generate the regenerative and developmental processes observed in the planaria.**

Hibernation

Hibernation is a state of inactivity and metabolic depression in warm-blooded animals such as squirrels, hamsters, hedgehogs, polar bears, and bats. Hibernation usually occurs in winter months characterized by a general lowering of metabolic rate, body temperature, breathing, and heart rate. Hibernation may last several days, weeks, or months, depending on the species.

Hibernation devastates the central nervous system of these animals. Their neurons shrink and thousands if not millions of vital connections between brain cells shrivel. Extensive pruning occurs in areas necessary for long-term memory such as the hippocampus.

A good example of a hibernating mammal is the arctic ground squirrel (FIG 6.3). Every September in Alaska and Siberia these squirrels retreat into burrows more than a meter beneath the tundra, curl up in nests built from grass, lichen, and caribou hair, and begin to hibernate. Their core body temperatures plummet, dipping below the freezing point of water. As a result, massive destruction of their cortical neurons takes place. Yet upon recuperation the arctic ground squirrel, as well as the majority of hibernating animals, demonstrate intact memory from their past by kin recognition, identification of familiar as compared to nonfamiliar animals, and retention of trained tasks.

In an Austrian study, ground squirrels were trained in summer to successfully accomplish two tasks: a spatial memory task in a maze and an operant task on a feeding machine. In spring, the same tasks were repeated. The hibernating group did not do as well as the control group in performing the learned tasks. On the other hand, the trained animals were successfully able to discriminate familiar from unfamiliar individuals in their group. It is unclear why the animals could demonstrate memory in one assay but not others, though the authors speculate it could be the result of the complexity of the task or the brain region responsible for the memories. This study does prove the persistence of social memories after considerable loss of neurons in the brain during hibernation.

In addition to lowering their metabolic rate, another way that hibernating animals survive their long state of torpor is by selectively initiating *autophagy*, a process that refers to cells literally consuming themselves. By removing the tissues the animal does not require while hibernating the animal saves energy for the remaining tissues.

When ground squirrels emerge in the spring after hibernating all winter, their guts have shrunk to about half their original weight, but their hearts remained unaffected because they kept beating though at a much slower rate. Guts and neurons are not the only things that shrink. When male ground squirrels wake up in the spring, they find their gonads shrunk to almost nothing. But relief is on the way. They soon regrow.

In the following study, from Germany this time, Alpine marmots were trained to jump on two boxes or to walk through a tube. When retested after six months of hibernation, their abilities were found to be unimpaired. The scientists conducting the study concluded that long-term memory is unaffected by hibernation in Alpine marmots.

Shrews are even smaller than marmots. While diminutive in size, when it comes to enduring the hardships of cold weather, they are biological giants. A new study from the Max Planck Institute used X-ray images to show that individual shrews decreased the size of their braincases in anticipation of winter by an average of 15.3 percent. Braincases then partially regrew in spring by 9.3 percent. The dramatic changes in skull and brain size apparently did not adversely affect their post-hibernation behavior.

Few scientists have conducted research on the memory retention capacities of bats that are also hibernating animals. Scientists at the Mammal Research Institute of the Polish Academy of Sciences undertook some truly interesting experiments. They trained bats to find food in one of three maze arms. After training, all bats performed 100 percent correctly. Then they went into hibernation. When they "woke up" the hibernated bats performed as before hibernation and as well as the non-hibernated controls.

The scientists concluded that bats benefit from an as yet unknown *neuroprotective* mechanism to prevent memory loss in the hibernated brain. Biochemical studies on the brains of frozen wood frogs have revealed various neuroprotective factors implicated in the promotion of tissue survival. While all these factors probably play a role in preserving a small collection of neurons that will form the scaffolding for the growth of new neurons after the animal "wakes," they cannot possibly be responsible for preservation of complex memories.

No matter how you look at it, **all these findings speak to the preservation of memory after hibernation.**

Metamorphosis

Holometabolous insects are insects that traverse four life stages from egg to larva to pupa to adult in the process known as *metamorphosis*. They are subject to extensive neurogenesis, pruning, and cell death in their brains. In spite of these radical changes in their cerebral cortices, it has been shown that memories from earlier stages of their existence survive the drastic reorganization of their nervous system when they reach adulthood.

These insects also experience major changes in body form, lifestyle, diet, and the use of particular sensory modalities. Is a maggot or caterpillar the same animal or different from the noisy blowfly or colorful butterfly that eventually emerges from the transitional pupal stage? Are you the same person you were when you were born? The same person as when you were conceived? In some ways, I would say, the same but also very different.

Most important, in spite of the changes our bodies undergo as we age, the memories of all our experiences from conception on, though not always accessible, persist etched into our embodied mind.

The adult moth's brain contains about one million nerve cells. In comparison, the human brain has more than one hundred billion. Yet there is a lot going on inside that moth's pinhead brain. A team of researchers at Tufts University studied learning in the *tobacco hornworm, Manduca sexta* (FIG 6.4), a species of moth. The researchers exposed the larvae of this species to the smell of ethyl acetate (EA) paired with a mild electric shock. When offered the choice of fresh air or EA-scented air in a Y choice apparatus naïve *fifth instar*[2] *caterpillars* (FIG 6.5) showed neither attraction nor aversion to the odor of EA. Larvae exposed to shock alone showed no attraction or aversion to EA. Larvae exposed to EA alone in the absence of shock showed no attraction or aversion to EA. However, the pairing of EA with electric shock (odor prior to shock) produced a significant aversion in fifth instar larvae with 78 percent of caterpillars choosing ambient air over EA.

The adult moths had retained what they learned as larvae, demonstrating a persistent and stable memory. Two possible mechanisms could explain such behavior. One, exposure of emerging adults to chemicals from the larval environment, or two, associative learning transferred to adulthood via maintenance of intact cells either in their brains or in other tissues. The researchers showed that the adult aversion did not result from carryover of chemicals from the larval environment, as neither applying odorants to naïve pupae nor washing the pupae of trained caterpillars resulted in a change in behavior. Evidently, the adult moths' **behavior represents true associative learning, not chemical legacy, and, as far as is known, provides the first definitive demonstration that associative memory survives metamorphosis in moths and butterflies.**

Yukihisa Matsumoto at Hokkaido University was similarly able to demonstrate the existence of long-lasting aversive memory in the hemimetabolous cricket, *Gryllus bimaculatus*, which retained an association between an odor and salt water for up to ten weeks. The learned preference was altered when at six weeks after the original training they were given "reversal training." The researchers concluded that crickets are capable of retaining olfactory memory practically for their lifetime and of easily rewriting it in accordance with new experience. If only humans were that smart.

Frogs are another species of animals that undergo metamorphosis. Peter G. Hepper, at Queen's University in Belfast, injected frog eggs with one of two substances, orange or citral, respectively. After hatching, these tadpoles preferred to feed on food containing the particular substances they were exposed to when they were still eggs. Even more surprisingly, after the tadpoles metamorphosed into frogs, they maintained the acquired odor preference. Preferences of tadpoles were also influenced "naturally" by odors present in the water surrounding the developing embryos.

In another series of experiments tadpoles that had been reared as embryos in orange-flavored water significantly preferred the orange side of the aquarium to the control side. Control tadpoles exhibited a preference for the water side of the aquarium. Hepper stated that his experiments demonstrated for the first time **"embryonic" learning in amphibians.**

After eggs of ringed salamanders were exposed to chemical cues from predators, post-hatching larvae showed reduced activity and greater shelter-seeking behavior. Working with wood frog larvae, researchers at Missouri State University conditioned them to chemical cues from unfamiliar predators. When the frogs reached adulthood, they responded with fear to the same cues. Similarly, larvae that had been exposed to neutral cues did not show these behaviors. Since embryonic experience is a good predictor of future conditions the organism will encounter, it follows that learning associated with exposure to negative stimuli during early development will be adaptive.

Finally, studies of other holometabolous animals, including beetles, fruit flies, ants, and parasitic wasps, have repeatedly and convincingly demonstrated that larval experience instructs adult behavior.

Neuroscientist May-Britt Moser at the Norwegian University of Science and Technology, one of the world's foremost scholars on the biology of memory, believes that brain cells of insects and other metamorphosing animals and those of higher animals use largely identical cellular mechanisms. Studies of insects and snails are decidedly relevant to understanding cognition, memory, and behavior in humans because as each species evolved there was no need for it to reinvent the wheel.

Summary

In this chapter we are addressing a very basic question: Can stable memories remain intact in animals that undergo massive loss and rearrangement of their cerebral neurons? There is the mystifying ability of planaria that can develop into "new" individuals from small portions of their "old" bodies and remember what they learned before they were cut into pieces. I align myself on this point with Tufts University biologists Shomrat and Levin, who proposed planarians as a key emerging model species for the mechanistic investigations of the encoding of specific memories in biological tissues. Moreover, this research will likely further the development of stem-cell-derived treatments of degenerative brain disorders in human adults.

Based on the above three classes of **animals in which memories survive drastic cellular turnover and rearrangement, it seems credible to conclude that memory, in addition to being stored in the brain, must also be encoded in other cells and tissues in the body. In other words, we are all endowed with both somatic and cognitive memory systems that mutually support each other.**

KEY TAKEAWAYS

- Retention of the learned response survives regeneration in planaria.
- Social memory does not seem to be affected by hibernation.

- In some cases of holometabolous insects, adult behavior has been shown to have originated from larval associative learning.

- Convincing data from planaria, insects, and mammals suggest that memories of learned behavior can survive drastic rearrangement and rebuilding of the brain.

CHAPTER SEVEN

THE WISDOM OF THE BODY

There is more wisdom in your body than in your deepest philosophies.

—Friedrich Nietzsche

Introduction

In the previous chapter we discussed how the planarian worm can develop into a full-fledged adult animal from just a tiny portion of another planarian. Similar, although less dramatic, is the transformation of bodies, including the brains of metamorphosing animals and hibernating animals in regard to retention of memories from their past. Humans are not that versatile. However, our bodies are perfectly capable to heal wounds or broken bones or even replace a complex structure such the liver. In order for healing to occur, cells, tissues, and organs somehow need to know what, when, and how to accomplish this. Without the ability of "remembering" the body's own structure, healing and regeneration of body tissues would not be possible.

As Marek Dudas, of the developmental biology program at Los Angeles' Saban Research Institute remarked, **"Tissue structural memory is distributed throughout all body structures and encoded directly within them,**

starting with DNA as [providing] the primary instructions for building all higher-order, more or less self-remembering systems." To put it more simply, repair of damaged tissues or organs is not top-down but bottom-up, not controlled by the brain but organized locally by the affected cells or assemblies of cells in body tissues and organs.

Harvard Medical School's Bessel van der Kolk, one of the foremost authorities on Post Traumatic Stress Disorder, wrote some years ago, **"Your body, believe it or not, remembers everything. Sounds, smells, touches, tastes. But the memory is not held in your mind, locked somewhere in the recesses of your brain. Instead, it's held in your body, all the way down at the cellular level."**

Body Memory

Twenty years ago, when I still played tennis in the summer and skied in the winter, a book by W. Timothy Gallwey called *The Inner Game of Tennis: The Classic Guide to the Mental Side of Peak Performance* became a best seller. Gallwey followed up with golf, music, etc. He had a hit on his hands. His advice for better performance: *overcome mental obstacles, improve concentration and reduce anxiety. Along with this, imagine as closely as you can how to hit a forehand, backhand or how to serve. Repeat this in your mind over and over again. In skiing, before you ski down the hill, think of where you will make the turns, which ski you will put your weight on as you make the turn, etc. The more you practice these things in your head, the better you will perform at the sport of your choice.* The muscles somehow learn from the mind how to respond in the desired way. Obviously, the connection between the mind and the body is very strong. But the reverse is less known but equally true.

In sports circles it has often been observed that fitness training in one's youth makes it easier to regain muscle mass and performance later in life, even after a long intervening period of inactivity. While in the past lasting effects of previous training were attributed to motor learning in the central nervous system, new studies show that they are a function of **muscle memory.**

A case in point is a Norwegian study that ascribes muscle memory as resulting from the peculiar properties of muscle cells. While most cells in the body have just one nucleus, muscle cells because of their large size have multiple nuclei. These nuclei store the genetic information for the cell. When a person uses their muscles, the muscle cells enlarge and develop more nuclei. If exercise or training is discontinued, the muscle cells shrink. What is exciting about this Norwegian research is the discovery that the extra nuclei remain long after training stops. Then, when one starts exercising again, the nuclei are still there, ready to support re-expansion of the muscles. The new muscle cell nuclei according to this study are more or less permanent once they are formed.

At the annual meeting of the American College of Sports Medicine in San Diego in 2017, researchers from Temple University presented preliminary results suggesting that, like muscle size and their multiple nuclei, an increase in mitochondria in muscles is equally important. As the mitochondria gained in an initial bout of training gradually disappear when one discontinues the training, the genes that control the formation of new mitochondria contained within the extra nuclei remain in the tissues as long as the organism is alive. Therefore, when a person starts training again after a long break, their muscle cells are already primed to begin producing more mitochondria.

Many functions of our bodies are controlled by internal biological or circadian clocks, which cycle daily and are synchronized with solar time.[1] These "clocks" regulate the sleep-wake cycle as well as the timing of human brain function, physiology, and behavior. They trigger the release of the hormone *melatonin* during sleep; favor the secretion of digestive enzymes at lunchtime; regulate body temperature, including blood pressure and heart rate; and keep us awake at the busiest moments of the day.

Biochemists at Bielefeld University have been researching the inner clock of plants for twenty years. They recently published a new study that shows that individual genes in plants, animals, and humans control the inner clock. Messenger molecules—messenger RNAs—are produced on these genes at a certain time of day. These molecules start the formation of clock proteins, which in turn reach their highest concentration at a fixed time of day. The clock proteins also ensure that other genes in the cell are active at the best possible time of day. They initiate different processes at certain times of the day: from opening and closing the petals in flowers to the sleep-wake rhythm in humans.

Scientists exploring the brain for answers to certain sleep disorders recently discovered that a circadian clock protein in the muscle—BMAL1—regulates the length and manner of sleep. Mice with higher levels of BMAL1 in their muscles recovered from sleep deprivation more quickly. Removing BMAL1 from the muscle severely disrupted normal sleep, leading to an increased need for sleep and a reduced ability to recover after sleep deprivation. The study demonstrates that at least one muscle protein, BMAL1, affects the brain. This is another excellent example of the operation of the body-to-brain feedback circuit.

Circadian studies are receiving increasing scientific recognition. In 2017 the Nobel Prize in medicine and physiology was awarded jointly to Jeffrey C. Hall (University of Maine), Michael Rosbash (Brandeis University), and Michael W. Young (Rockefeller University) "for their discoveries of molecular mechanisms controlling the circadian rhythm." By examining the internal workings of fruit flies, the investigators determined that the gene they were analyzing encoded a protein that accumulated in cells at night, and then degraded during the day (we are back to epigenetics here).

Light is now recognized as the principal circadian synchronizer in humans. In sighted people, the intrinsic circadian period is approximately 24.2 hours long. It is the same in blind individuals. Therefore, it is not through our eyes but rather by way of the cells in our skin that light or lack of it registers. I am glad the circadian system is receiving long overdue scientific recognition because it illustrates another aspect of the Embodied Mind.

Of interest, though not totally relevant, is the finding that the circadian rhythms of melatonin and body temperature are set to different times in women than in men, even when the women and men maintain nearly identical and consistent bedtimes and wake times. Women tend to wake up earlier than men and exhibit a greater preference for morning activities than men. The neurobiological mechanism underlying this sex difference in circadian alignment is unknown.

I think it is important to point out that when cells in our body multiply, they follow a program known as the *cell cycle*. In a normal cell cycle two daughter cells are produced by cell division at the end of the cycle. Many cancers involve a dysfunctional or hyperactive cell cycle, which allows the tumor cells to multiply uncontrollably. In the past it was assumed that the cell cycle was independent of the circadian clock. However, a study published in the December 2017 issue of the journal *PLOS Biology* showed that a protein called RAS, known to control the cell cycle, also controls the circadian clock. Their work highlights the importance of the circadian clock as modulator of cell function and further reinforces the important role it plays in cancer prevention.

When one walks or drives it is body memory that does the primary computations, with little or no input from higher brain centers. One does not have to think, "Lift left foot, put left foot down, lift right foot, etc." We shift into what could best be described as automatic pilot.

A familiar example of body memory at work is the healing of injured tissues. When tissues are damaged, organic junk has to be removed from the site of injury, and new components must be delivered or recreated there. Everything must happen in an orchestrated manner and involves genes, cell signaling, matrix interactions, functional vasculature, and neurohormonal regulation circuits, to name just the most important elements. The tissues and cells always perform their reparative tasks using a predetermined set of biologic functions, and under the influence of microenvironmental conditions. It needs to be emphasized that the guidance for repair originates not in the brain but locally, in the damaged tissues and organs, in concert with other structures near and far. The body performs this task under the radar, so to speak, mysteriously and efficiently outside of our awareness.

As we learned in the chapter on the immune system, its primary function is to maintain memories of past inflammations to mount faster responses to recurrent infections. However, scientists in many parts of the world suspected that other types of long-lived cells might similarly remember inflammation. Skin was a logical place to investigate: as the body's outer coating, it is exposed to frequent attacks.

Now, researchers from Rockefeller University showed in experiments with mice that wounds closed more than twice as fast in skin that had previously experienced inflammation than in skin that had not, even if that initial inflammatory experience had occurred as long as six months earlier, the equivalent of about fifteen years for a human. Healing sped up, the team determined, because the inflammation-experienced *stem cells* that remained embedded in the deeper layers of skin were better at moving into the wound to repair the breach.

Signals from our five senses inform the brain about our external environment. If a little girl touches a hot stove she will most probably cry out from the pain and reflexively withdraw her hand. The memory of the event will cause her to avoid touching a hot object in the same way again. Sensory signals from the skin, eyes, and other organs create memories in the brain. They also leave behind imprints in the cells, assemblies of cells, and the neural networks that transmitted these signals. The more senses are activated by a stimulus, the better we recall the event because additional cells in the body, along with the brain, have participated in processing and remembering the experience.

A common ordeal for many people, especially the elderly, is chronic pain. Scientists tell us that this highly distressing condition induces specific changes in neurons and microglia. It seems that small and often seemingly innocuous injuries leave molecular "footprints" behind which add up to more lasting damage, and ultimately chronic pain. In healthy people, cells' housekeeping protein systems replace and restore the majority of their content every few weeks. In sufferers of chronic pain, however, damaged copies of essential proteins replace the healthy ones. Why do these changes persist long after the precipitating injury has healed? A new study from King's College London offers at least a partial answer.

Examining the spinal cord immune response in neuropathic pain, they isolated resident microglia for genome-wide RNA sequencing (RNA-seq) and epigenetic profiling. Their data provide evidence that peripheral nerve injury changes microglial enhancers, what we called switches in the chapter on genetics, a process that could maintain these cells in an abnormal, maladaptive state over long periods. Simply put, the scientists found for the first time that chronic pain may be caused by and persists on account of epigenetic changes in microglia.

Chronic pain is also frequently present in people suffering of a *phantom limb*. A phantom limb is the sensation that a person with an amputated or missing limb, like the arm, experiences as if it was still attached to the body. Commonly, in addition to pain, some people register other sensations in their phantom limb such as tingling, cramping, heat, and cold.

Theories of proposed mechanisms to explain phantom sensations have changed in the past few years from believing that it was psychological in origin to hypothesizing that it is caused by peripheral and central nervous tissue changes. More recently, the role of mirror neurons in the brain has been proposed in the generation of phantom pain. In 2016 a team at Oxford's Hand and Brain Lab used an ultra-high-power MRI scanner to compare brain activity in two people who had lost their left hand through amputation but who still experienced vivid phantom sensations with eleven people who retained both hands and were right-handed. Each person was asked to move individual fingers on their left hand. The scientists involved in the study remarked that while the brain does carry out reorganization when sensory inputs are lost, it does not erase the original function of a brain area.

In view of the above discussion, I think it is safe to assume that numerous mechanisms are responsible for the existence of phantom limb. One mechanism that has not been investigated, is the one I propose here; namely, that the cells and tissues that once had contact with the removed part "remember" it, just as the sensory brain area that originally served that function and continue to send signals to the brain as if the part was still there. It seems that they do so in spite of the fact that the person so afflicted knows full well that the part is missing. In other words, there is a disconnect between the cognitive centers of the brain (our conscious self) and the community of cells, part of our embodied brain, which acts as if the limb or organ was still present.

I invite you to consider this recollection of one of my friends: *"My mom has severe dementia. She doesn't seem to know anything and very rarely speaks. She had eight children and is legally blind. Sometime ago I brought her a doll. I thought it would comfort her. When I gave her the doll she 'felt' around the doll because she cannot see. She then turned the doll and positioned it as you would a baby. She sat contently holding the baby doll. I feel like that was body memory because my mom's life revolved around raising children and grandchildren."*

For a moment, and perhaps for the first time in months, my friend's mother was centered, oriented not just to the doll in her arms and to her daughter, but to her own identity. A moment that unified her unconscious physical habits and consciously remembered experiences with her present intentions, bodily sensations and deeply felt emotion. My friend's account highlights the enduring effect of body memory. Even after the brain had lost its cognitive functions, even after a person had become deaf and blind, the body still remembers.

William James—who shaped our understanding of the psychology of habit as far back as 1884 in an essay titled "What Is an Emotion?"—made a persuasive case for "how much our mental life is knit up with our corporeal frame." With an eye to the intense interplay between our bodies and our minds Marc Wittmann, outstanding German psychologist and chronobiologist, wrote in 2016, "The brain does not simply represent the world in a disembodied way as an intellectual construct, but rather the organism interacts as a whole with the environment . . . and social interaction with other people. **Our mind is body-bound."**

This seems like a good place to resurrect Wilhelm Reich, a student of Sigmund Freud, who broke away from Freud because he wanted to focus more on the body and less on the mind. Reich advanced the term *character armor*, which he defined as the bodily expression of the conflict between our basic human needs, desires, and authentic feelings with the unnatural attitudes and conditioning imposed upon us by family and society. Armoring is a way of protecting ourselves from the pain of not expressing those parts of ourselves that we have learned to hide.

Years of unresolved, emotionally painful memories create hardened tissues that fixate movement and pull the body out of its natural alignment. The longer this state persists, the more it becomes part of our body armor. Body armor is evident in people of all ages, including children and the elderly. A common symptom of people with body armor is fatigue experienced because it takes energy to suppress body memory, especially unresolved emotional energy.

Health professionals such as osteopaths, chiropractors, craniosacral therapists, massage therapists, and others working with their clients' bodies generally subscribe to the belief that rather than festering anxiety, it is physical, sexual, or emotional trauma that has created a localized, compressed area or areas of foreign, disorganized energy in the body, walled off from consciousness. Working on these areas will often uncover and liberate traumatic memories and/or free the person of chronic pain. I think both theories are correct. With some people body armor develops as a result of an unresolved conflict while in others it may be due to a traumatic experience.

When the experience is life-threatening or takes place over a protracted period and is excruciatingly painful, either physically or psychologically or both, the person may experience a total psychological breakdown. All of us have limits as to what we can tolerate in terms of physical, sexual, or emotional abuse. When that limit is breached the pain gets pushed out of consciousness and buried deep in the unconscious. The survivor's mind blocks out the pain and "forgets" what happened until many years later, when some chance occurrence or entering psychotherapy brings it back, sometimes only partially, other times in its entirety. There is a very natural resistance to unearthing old traumas. Remembering promises to be distressing. On the other hand, the body is eager to tell its story because it wants to unburden itself of its painful and often shameful secrets.

People with chronic unremitting trauma are diagnosed as suffering of *post-traumatic stress disorder* (PTSD), a condition characterized by persistently high arousal, recurrent, involuntary, and intrusive memories of the traumatic event, memory loss for other parts of that event, lack of ability to concentrate, impairment of social functioning, and feelings of detachment or estrangement from others. The effects of the trauma persist because the traumatic event is locked into the cellular structures of the body. PTSD is another example of body memory.

In the case of people with PTSD a touch, smell, sound, even certain types of weather can trigger and bring the past suddenly into the present again. The writer Aharon Appelfeld put it this way in his memoirs:

"Everything that happened at that time has left its mark in the cells of my body. Not in my memory. The body's cells seem to remember better than the memory, which is intended for this. For years after the war, I did not walk in the middle of the pavement or path, but always close to the wall, always in the shade, always in a hurry like someone fleeing. (. . .) Sometimes it is enough to smell food, to feel dampness in my shoes or hear a sudden noise to bring me back to the war (. . .) The war sits in all my bones."

In the above example, it is not a particular episode, but an entire segment of a person's life that has left its mark on the body, more deeply and permanently than any autobiographic memory could ever achieve. Traumatized persons react to reminders of the trauma with emergency responses that were relevant at the time of the original threat, but unfortunately are usually inapropriate and ego alien in the current situation.

Interestingly, when Aharon Appelfeld said, "*The war sits in all my bones,*" what he probably meant metaphorically is actually being proven to be scientifically true. According to geneticists from Columbia University, the acute stress response in bony vertebrates is not possible without *osteocalcin*. Osteocalcin is a hormone made in bones. It regulates a large and continually growing number of physiological functions and developmental processes. Numerous studies have found that it helps regulate metabolism, fertility, brain development, and muscle function. It improves memory and lowers anxiety. It is linked to the biology of aging.

James Herman, a neuroscientist at the University of Cincinnati, commenting on this research, said, "I think what that means is that the way we currently understand stress is too simplistic." Herman adds that chemical messengers from other parts of the body may also play a role in the stress response. Experiments from his own lab have identified a potential stress-related role for signals secreted from fat. The fat-to-brain signal may be mediated by neuronal mechanisms, release of *adipokines,* or increased lipolysis. These studies support a link between stress biology and energy metabolism, a connection that has clear relevance for numerous disease states and their comorbidities (the simultaneous presence of two or more diseases or medical conditions in a patient).

Researchers from Columbia University and the Sorbonne pointed out that what this research **illustrates is a classic and fundamental principle of physiology: no organ, including the brain, is an island unto itself.**

The Good News

Kristian Gundersen and his colleagues have observed that the ability to generate new nuclei in muscles appears to decline with age. That suggests that getting as fit as you can while young will provide long-lasting benefits as you age, making your muscles more responsive to exercise.

Today a vast literature exists on people's responses to extreme experiences, such as combat trauma, rape, kidnapping, natural disasters, accidents, and torture. We most likely can never totally let go of these fearful memories. We can, however, diminish their impact on our daily lives and gradually learn to replace the negative judgments we made about ourselves following the traumatic experiences with an honoring of our strength of character to have survived and overcome them.

Body-centered therapy, also referred to as somatic psychotherapy, is a therapeutic approach that recognizes the intimate relationship between the human body and the psychological well-being of a person. It was first introduced by Wilhelm Reich, who used a combination of breath and intensive body work to release the contractions in muscles he felt were abnormaly constricted as a result of past stressful and traumatic experiences. His intent was to gain access to and free these painful memories and emotions that had led to body armoring.

The story told by the "somatic narrative"—gesture, posture, movement, the patterns of stress, and intonation in speech—are arguably more significant than the story told by words. Body-centered psychotherapy makes an end run around words and heads for the body. This approach has proven beneficial in treating chronic pain and PTSD by helping people to feel the experience directly in the body, circumventing intellectualizations and obsessive thoughts. In many instances (no therapeutic approach works for everyone) this is all that is needed to activate the body's natural relaxation and restoration functions and bring about healing. Of course, like all good therapy, to achieve full remission of symptoms takes time.

Massage therapy, along with various methods of psychological talk therapy, can be very helpful. Other modalities of healing the body are shiatsu, therapeutic touch, cranial-sacral therapy, polarity therapy, trigger point massage, rolfing, and Feldenkrais movement reeducation. As with all psychotherapy, one needs to find a therapist who is knowledgeable, skilled, and caring.

A phenomenon that has received little attention is what I call *chronopsychology*. The grandmother of one of my patients gave birth to the patient's mother, Judy, at age eighteen. Judy gave birth to Annie, my patient, and also at eighteen Annie gave birth to her daughter, Beth. You will be pleased to know that Beth broke this remarkable unconscious pattern. She had a son at twenty-two. Australian pediatrician Averil Earnshaw wrote a book on this subject. In it she says, "I write of neglected and undigested experiences in families, which erupt as repeat performances, at crucial ages in family lives . . . I put forward observations that these eruptions affect us at the very same ages that they were 'buried' in our parents' lives."

Earnshaw discusses many incidents of what she calls *Family Time* from her own practice and adds sixty-one case histories of famous people from Jane Austen to Virginia Woolf that demonstrate the effects of chronological body memory on their lives.

Please keep in mind that a person's belief in their own agency, degree of self-esteem, their approach to life, be it positive or negative—all these factors will have a huge effect on tissue healing and health in general.

Summary

The physical body as well as the brain retains a memory of all our experiences. If our cerebral and somatic memories did not exist and respond collectively, we would have to relearn every day the basic tasks of living, like brushing our teeth or opening a can of tuna.

Healing and repair of damaged tissues is triggered by reactions of these tissues. In other words, on a local level free of central command, read: brain. This localized reaction is followed by the operation of complex interactions between all structure-maintaining systems consisting of biomolecules, genes, subcellular and extracellular structures, and signaling networks, together with all related vascular, hormonal, and neural mechanisms.

Imagine for a moment that an old and famous theater in Boston is badly in need of repairs. The municipal government, with a budget allocated for such matters, will initiate the restoration by engaging a local architect and appointing a site manager who in turn will hire the necessary work crew. The state legislature may decide to contribute a certain sum of money, and with any luck, Congress may also contribute. In addition, there may be nonprofit organizations and even private individuals who decide to chip in. While the city of Boston is in charge, many other levels of government and organizations are also involved. You get the picture.

It is the existence in our bodies of an intricate, unified, multilevel, homeostatic, cellular memory system that allows us to be fully functional human beings.

KEY TAKEAWAYS

- Through the repetition of experiences, a habit is formed: well-practiced motion sequences become implicit bodily knowledge and skill.

- Experiences, especially ones that carry a strong emotional charge and/or have involved multiple sensory organs, such as the eyes, ears, or skin, tend to be more strongly anchored in body memory than experiences with less emotional charge or perceived by only one or two sensory modalities.

- Local cells and tissues originate and control repair as they gradually call on other systems to support their work.

- Injuries that cause inflammation are remembered by stem cells in the skin.

- Traumas, whether physical or psychological, are locked in the whole body. Therefore, complete healing may benefit from body-oriented psychotherapy.

- On occasion, our bodies speak loudly about things we would rather not hear. That is the time to pause and listen.

FIG 1.1 A female zebrafish. *Credit: Soulkeeper (Own work) [Public domain], via Wikimedia Commons.*

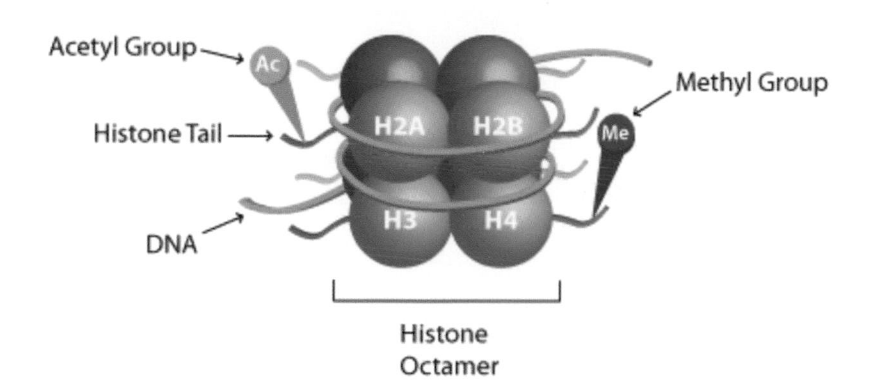

FIG 1.2 Histone. *Credit: EpiGentek.*

FIG 1.3 Human Early Learning Partnership. The Epigenome—Regulating Gene Activity [light switch on/dimmer/off]. *Credit: Dr. Michael S. Kobor, Human Early Learning Partnership. DNA Methylation—Active Gene, Silenced Gene [light switch on/off]. Vancouver, BC: University of British Columbia, School of Population and Public Health; 2014.*

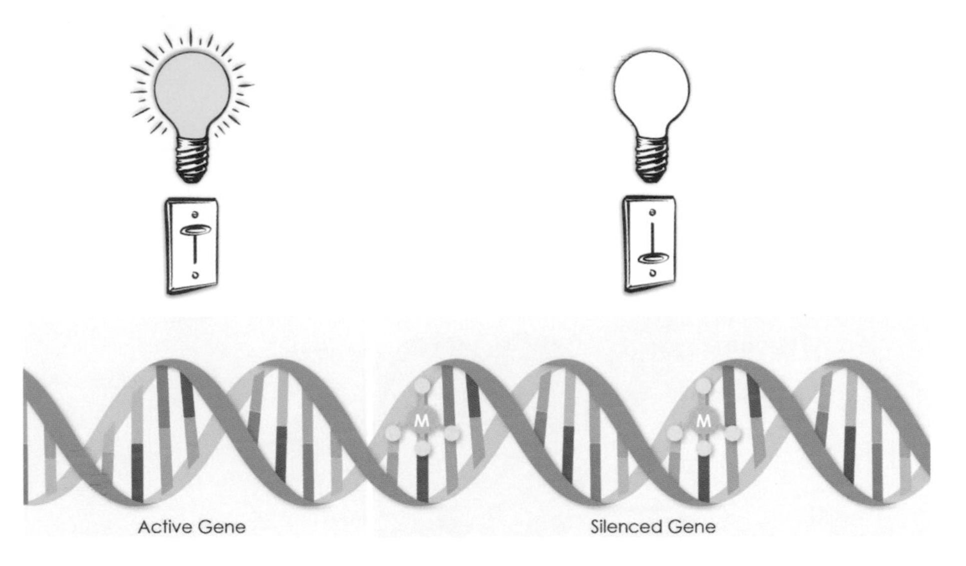

FIG 1.4 Human Early Learning Partnership. DNA Methylation—Active Gene, Silenced Gene [light switch on/off]. *Credit: Dr. Michael S. Kobor, Human Early Learning Partnership. DNA Methylation—Active Gene, Silenced Gene [light switch on/off]. Vancouver, BC: University of British Columbia, School of Population and Public Health; 2014.*

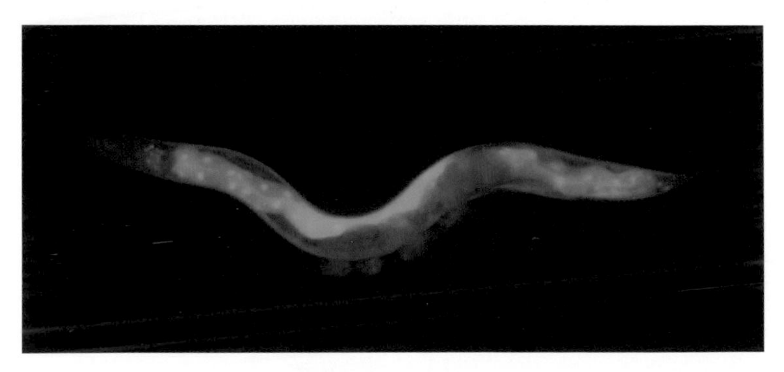

FIG 1.5 C. elegans.
Credit: Kapahi Lab, Buck Institute for Research on Aging, Novato, CA.

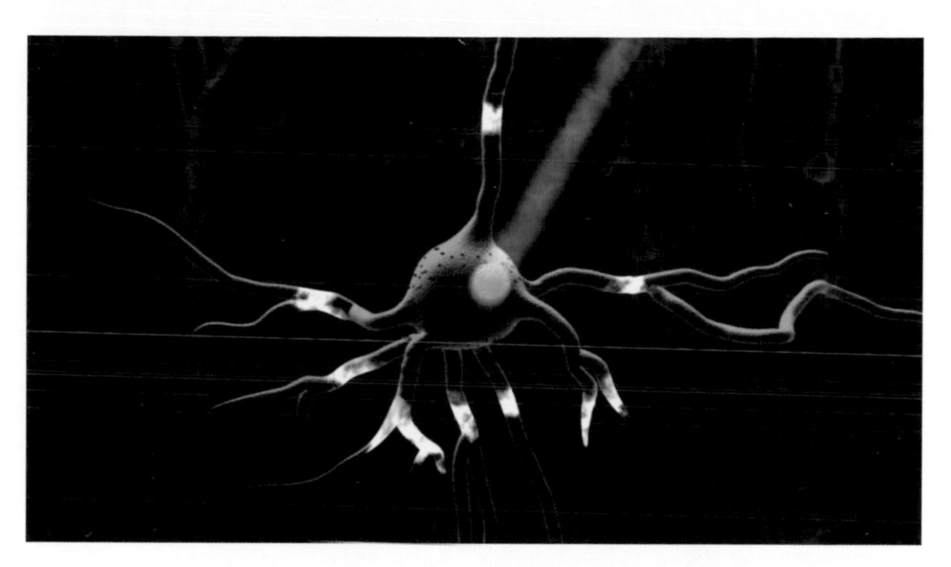

FIG 2.1 A neuron with one axon and several dendrites. *Credit: McGovern Institute for Brain Research at MIT.*

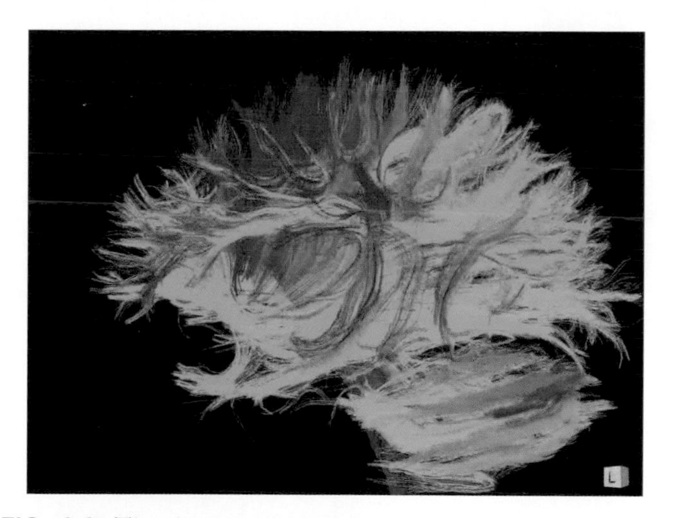

FIG 2.2 The human connectome. *Credit: jgmarcelino from Newcastle upon Tyne, UK—Webs'r'us. Uploaded by CFCF, CC BY-SA 2.0, https://commons.wikimedia.org/w/index.php?curid=31126898.*

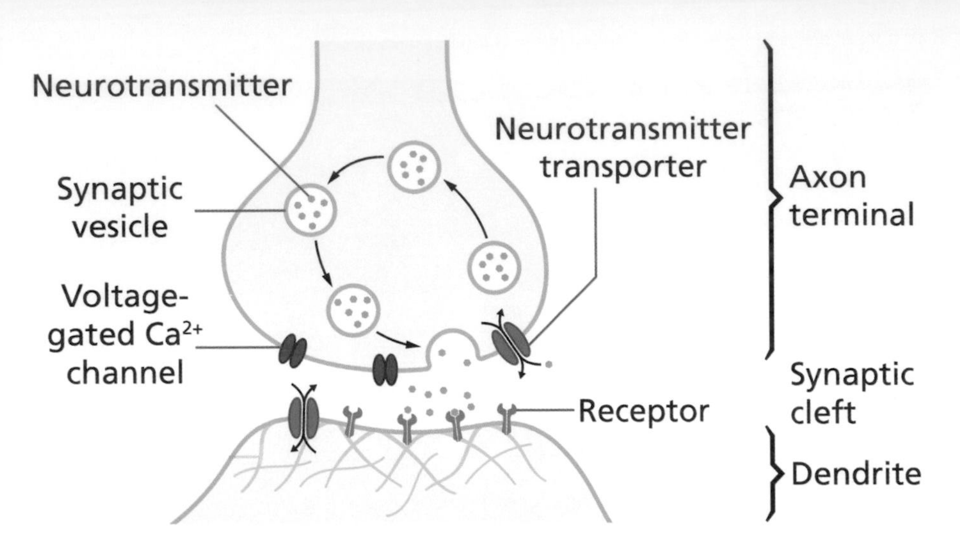

FIG 2.3 A synapse or junction. *Credit: Thomas Splettstoesser (www.scistyle.com)—Own work, CC BY-SA 4.0, https://commons.wikimedia.org/w/index.php?curid=41349612.*

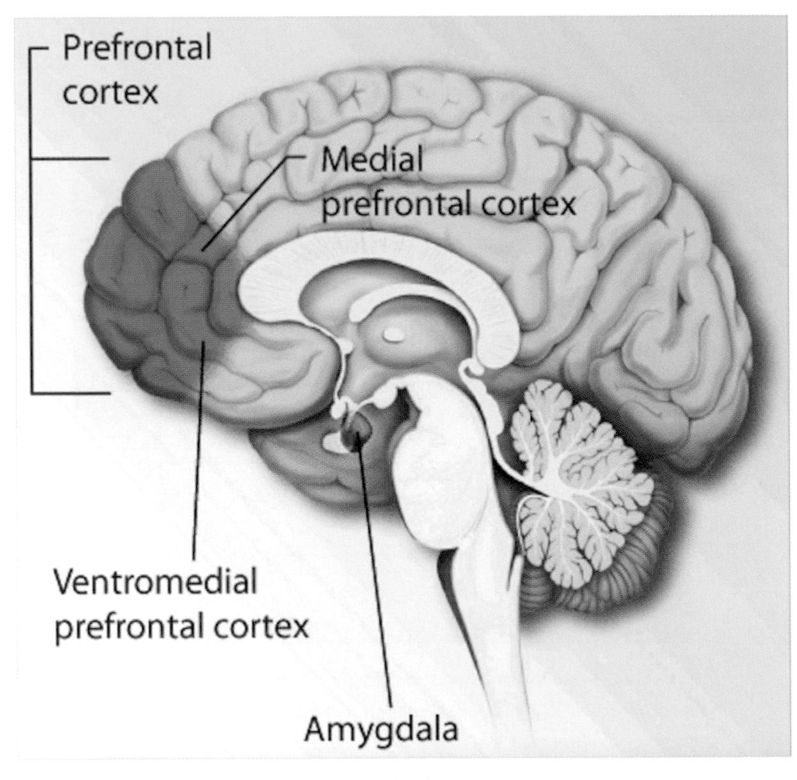

FIG 2.4 Cerebral cortex.
Credit: The National Institute of Mental Health (NIMH) [Public domain], via Wikimedia Commons.

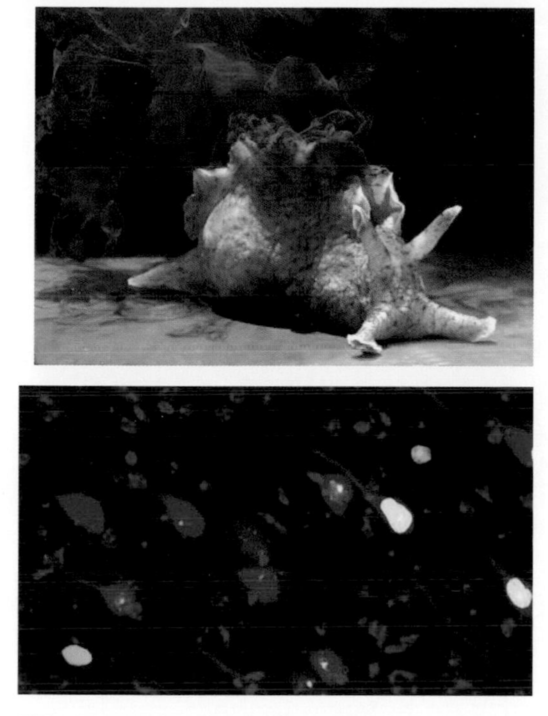

FIG 2.5 Aplysia californica. *Credit: Genny Anderson—http://marinebio.net/marinescience/03ecology/tptre.htm, CC BY-SA 4.0, https://commons.wikimedia.org/w/index.php?curid=264835.*

FIG 2.6 Memory engram cells (green and red), which are crucial for permanent memory storage in the prefrontal cortex. *Credit: Takashi Kitamura, Tonegawa Laboratory.*

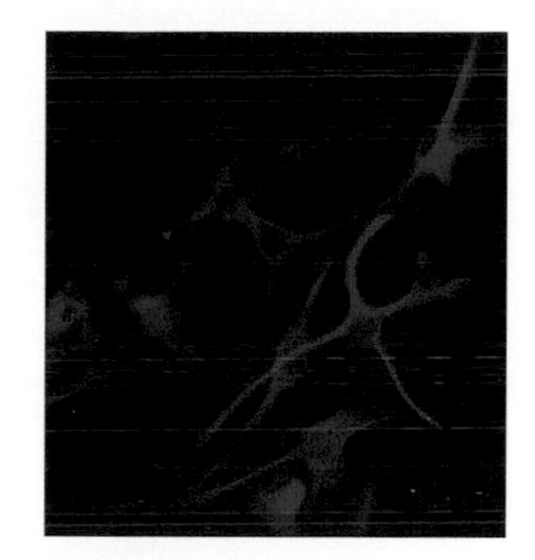

FIG 2.7 Astrocytes. From Fields, R.D., and Stevens-Graham, B. (2002). New Insights into Neuron-Glia Communication. *Science* 298: 556–562. *Credit: Kelly Fields.*

FIG 2.8 Amphioctopus marginatus hiding between two shells.
Credit: Nick Hobgood CC BY-SA 3.0, https://creativecommons.org/licenses/by-sa/3.0.

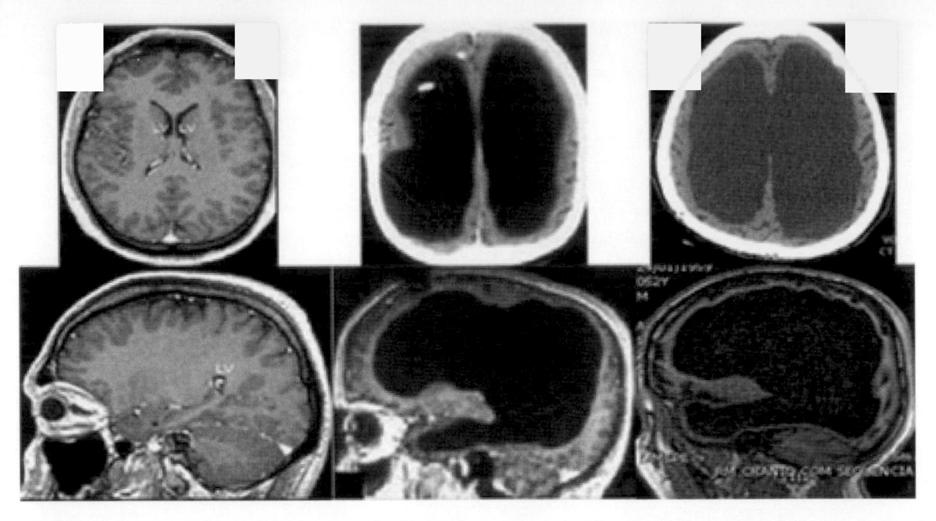

FIG 2.9 Normal adult appearance (left). Enlarged ventricles (middle and right). *Reproduced under Creative Commons License from Frontiers in Human Neuroscience (Forsdyke 2014).*

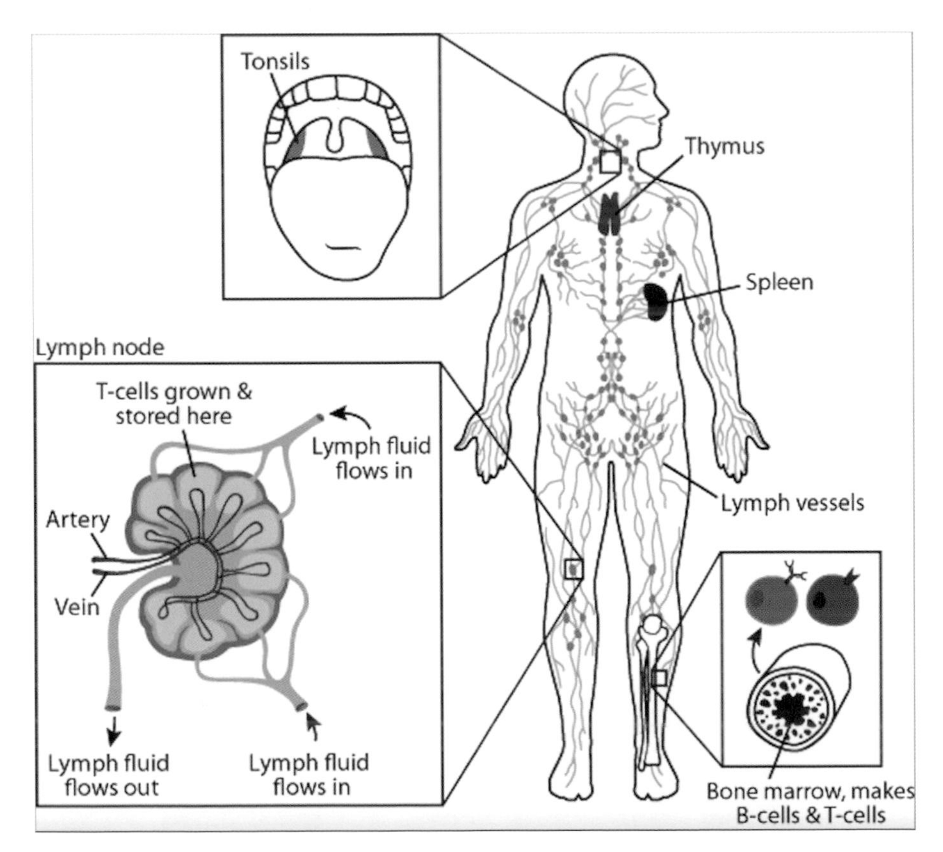

FIG 3.1 The lymph system. *Credit: © Arizona Board of Regents / ASU Ask A Biologist. CC BY-SA 3.0, https://askabiologist.asu.edu/t-cell.*

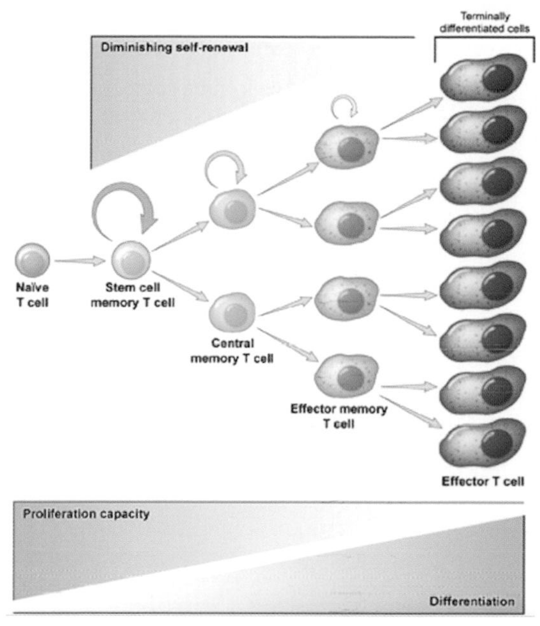

FIG 3.2 Memory T cells. *Credit: The National Cancer Institute's Center for Cancer Research.*

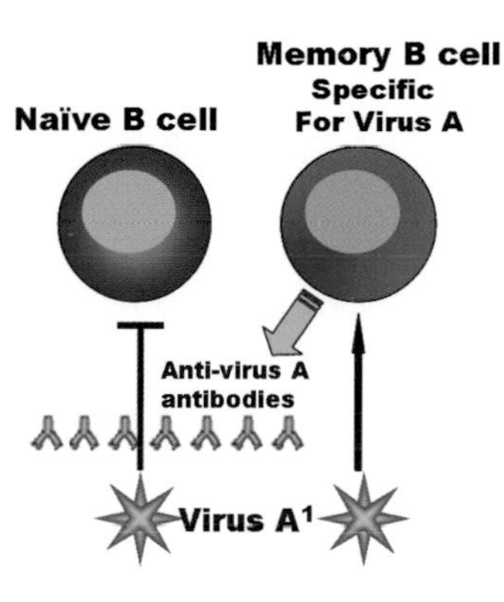

FIG 3.3 Memory B cell. *Original antigenic sin, also known as the Hoskins Effect (Endnote 3). Credit: CC BY-SA 3.0 (http://creativecommons.org/licenses/by-sa/3.0/), via Wikimedia Commons.*

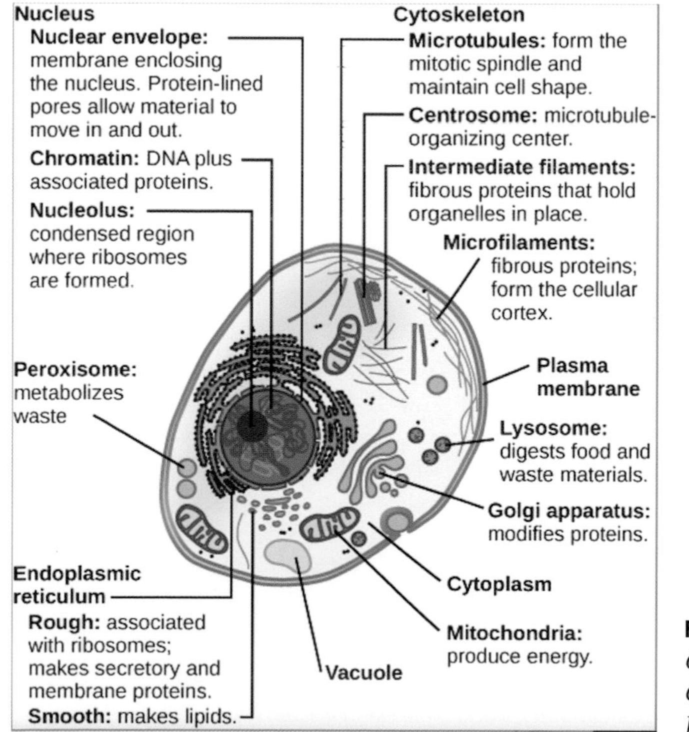

FIG 4.1 A human cell. *Credit: CNX OpenStax CC BY-SA 4.0, via Wikimedia Commons.*

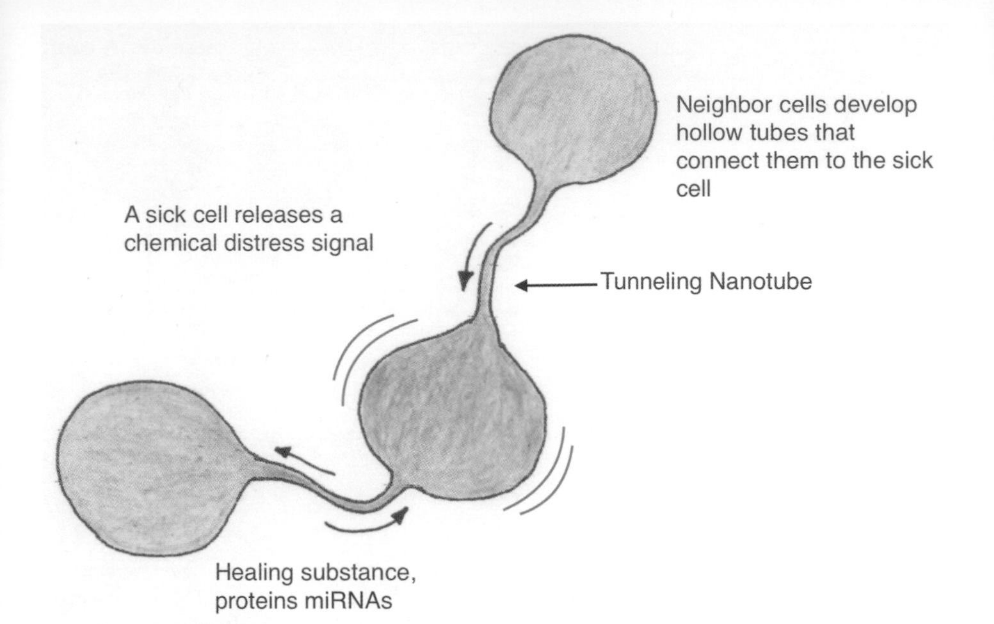

FIG 4.2 Tunneling nanotubes. *Credit: Bob Tadman.*

FIG 4.3 Muybridge racehorse gallop. Credit: *Photos taken by Eadweard Muybridge (d. 1904). Edit by User:Waugsberg [Public domain], via Wikimedia Commons.*

FIG 4.4 Matryoshka doll. *Credit: Fanghong CC BY-SA 3.0, via Wikimedia Commons.*

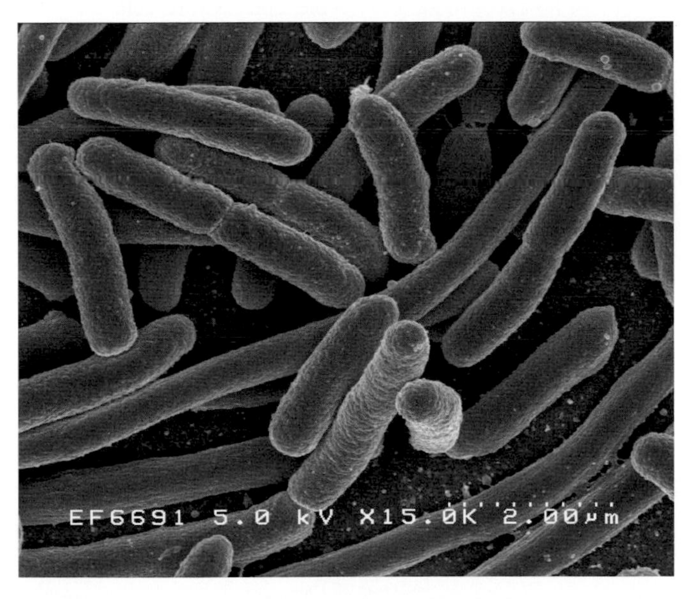

FIG 5.1 Escherichia coli: Scanning electron micrograph of Escherichia coli, grown in culture and adhered to a cover slip. *Credit: Rocky Mountain Laboratories, NIAID, NIH [Public domain], via Wikimedia Commons.*

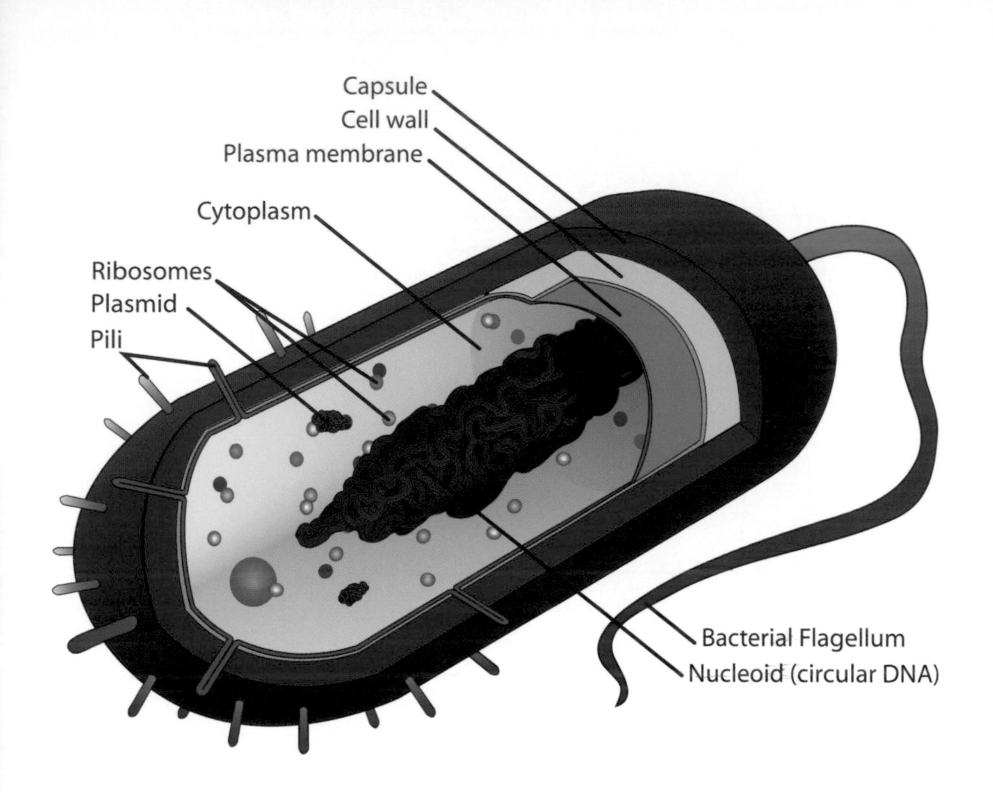

FIG 5.2 Structure of a typical prokaryotic cell.
Credit: Mariana Ruiz Villarreal [Public domain], via Wikimedia Commons.

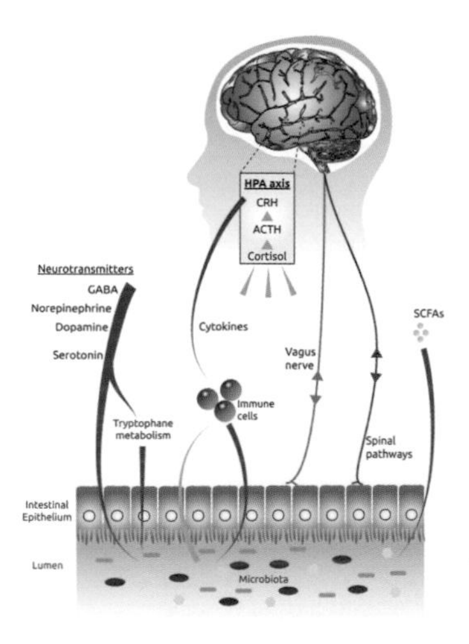

FIG 5.3 The multiple bidirectional routes of communication between the brain and the gut microbiota. *Credit: Dinan, T. G., Stilling, R. M., Stanton, C., & Cryan, J. F. (2015). Collective unconscious: how gut microbes shape human behavior.* Journal of Psychiatric Research, *63, 1–9.*

FIG 5.4 Plasmodial slime mold, Fuligo septica.
Credit: Scot Nelson, Attribution-ShareAlike 2.0 Generic CC BY-SA 2.0, via Flickr.

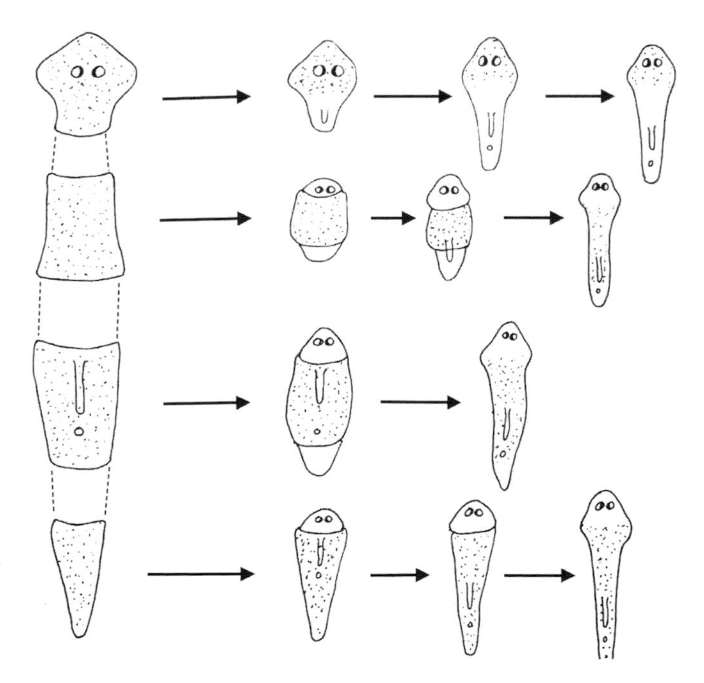

FIG 6.1 Decapitation and regeneration. Illustration of worm regeneration sequence.
Credit: Bob Tadman.

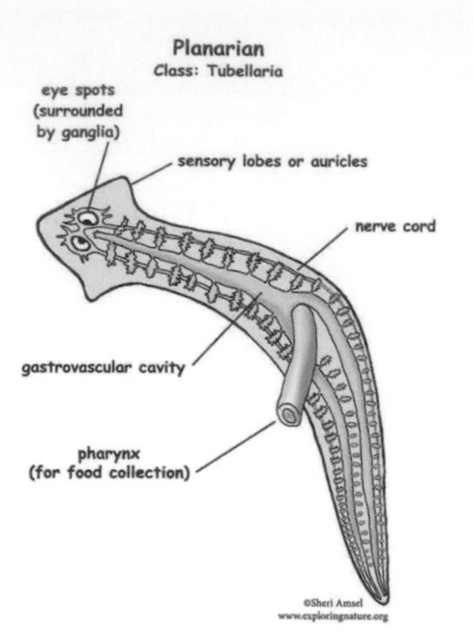

FIG 6.2 Planarian anatomy. *Credit: ©Sheri Amsel–Exploringnature.org.*

FIG 6.3 Arctic ground squirrel. *Credit: Public domain, via Wikimedia Commons.*

FIG 6.4 Tobacco hornworm pupa. *Credit: By Daniel Schwen CC BY-SA 4.0, via Wikimedia Commons.*

FIG 6.5 Tobacco moth (Manduca sexta). *Credit: Kugamazog—CC BY-SA 2.5-2.0-1.0, via Wikimedia Common*

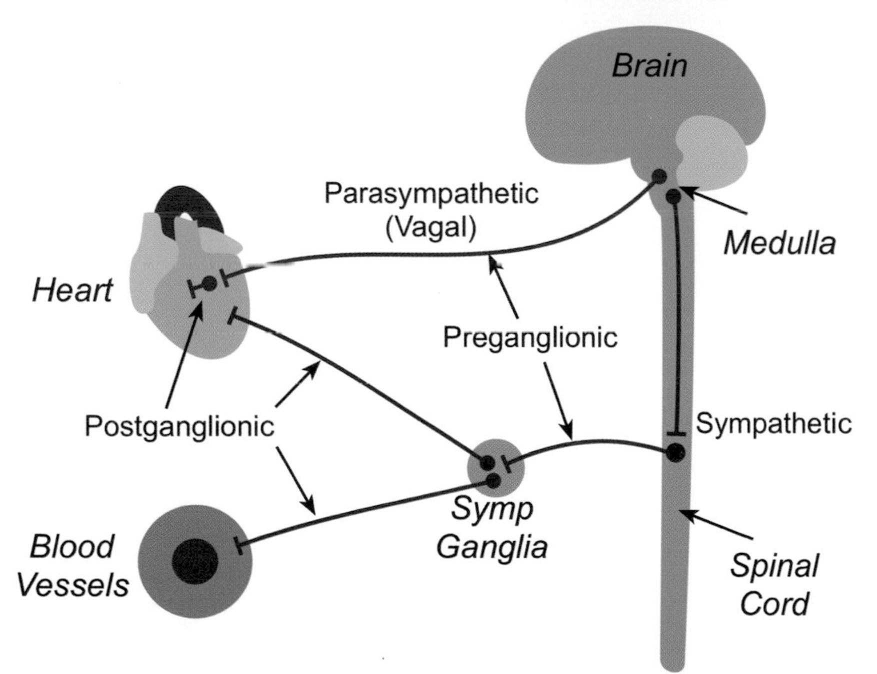

FIG 8.1 Autonomic innervation of the heart and vasculature. *Credit: Richard E. Klabunde, Ph.D. Marian University College of Osteopathic Medicine.*

FIG 9.1a 3 months after start of the experiment. *Credit: Bob Tadman.*

FIG 9.1b 6 months after the start of the experiment. *Credit: Bob Tadman.*

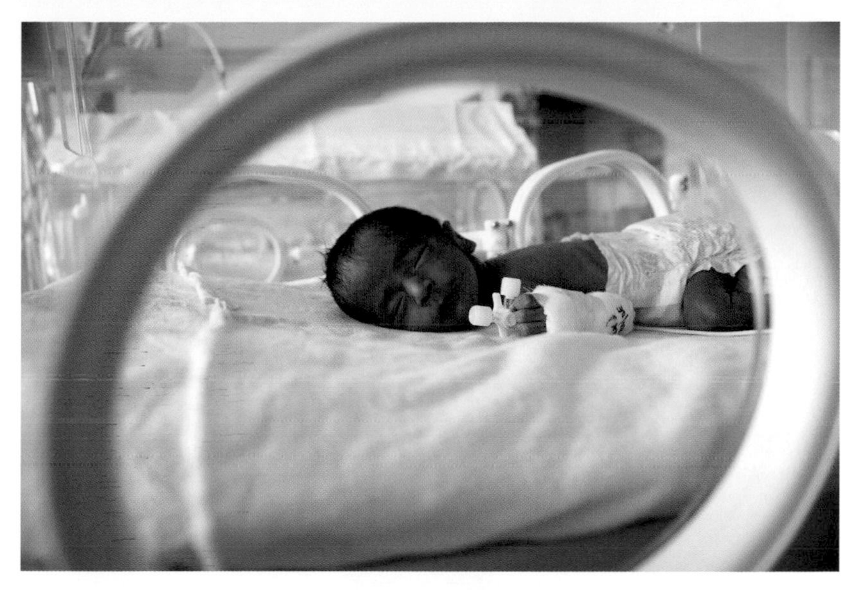

FIG 9.2 Newborn in NICU. *Credit: Creative Commons CC0.*

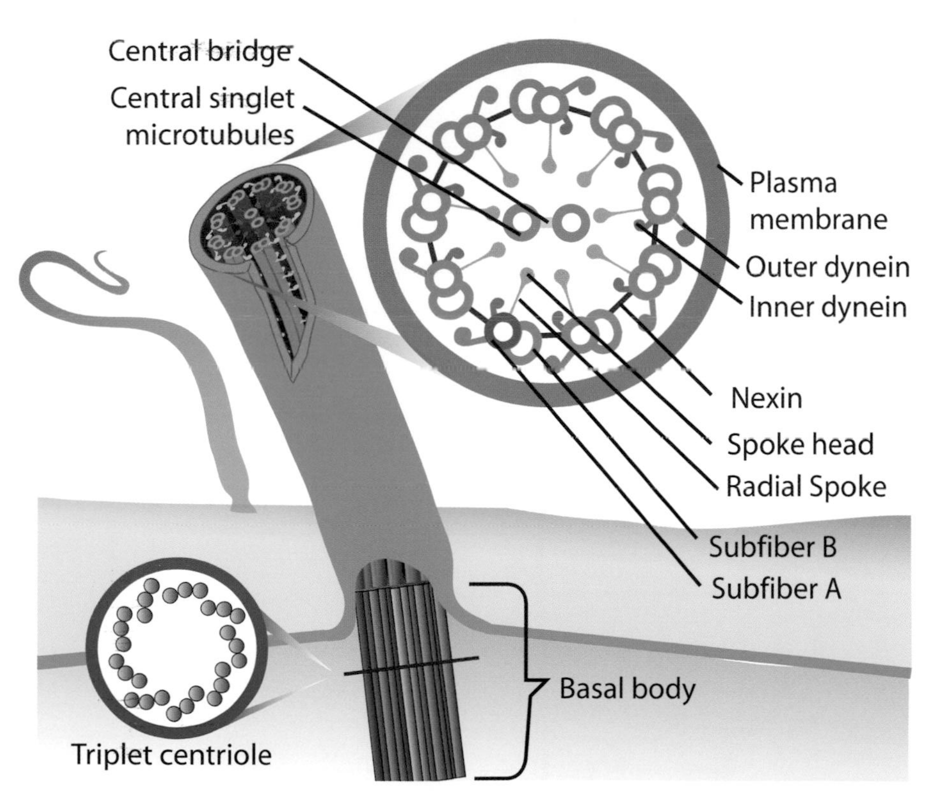

FIG 9.3 Cilia of paramecium showing microtubules.
Credit: LadyofHats [Public domain], from Wikimedia Commons.

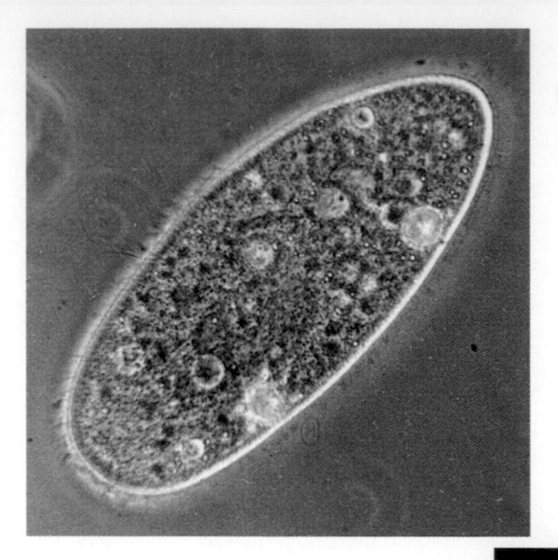

FIG 9.4 Paramecium.
Credit: Barfooz CC BY-SA 3.0, via Wikimedia Commons.

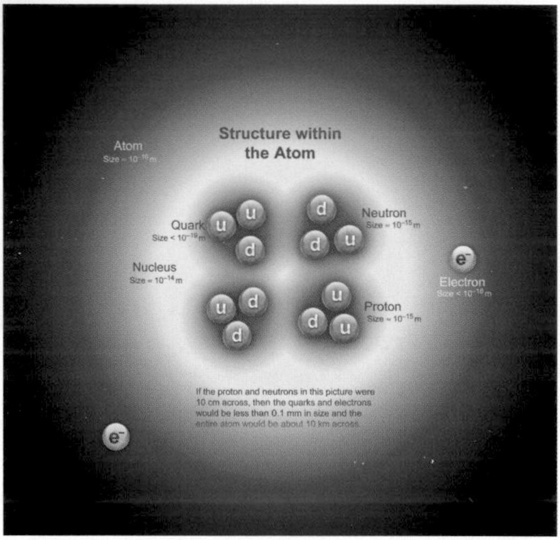

FIG 9.5 Image of an atom. *Credit: Contemporary Physics Education Project.*

FIG 9.6 Entangled particles. *Credit: Katie McKissik.*

CHAPTER EIGHT

HEART TRANSPLANTS: PERSONALITY TRANSPLANTS?

Sometimes the heart sees what is invisible to the eye.

—H. Jackson Brown Jr.

A loving heart is the beginning of all knowledge.

—Thomas Carlyle

It is only with the heart that one can see rightly; what is essential is invisible to the eye.

—Antoine de Saint-Exupéry

Nobody has ever measured, not even poets, how much the heart can hold.

—Zelda Fitzgerald

Your heart is full of fertile seeds, waiting to sprout.

—Morihei Ueshiba

The human heart has hidden treasures, In secret kept, in silence sealed; The thoughts, the hopes, the dreams, the pleasures, Whose charms were broken if revealed.

—Charlotte Brontë

A loving heart is the truest wisdom.

—Charles Dickens

Put your heart, mind, and soul into even your smallest acts. This is the secret of success.

—Swami Sivananda

Only do what your heart tells you.

—Princess Diana

Whatever your heart clings to and confides in, that is really your God.

—Martin Luther

The heart has its reasons of which reason knows nothing.

—Blaise Pascal

Common Heart Expressions:

Follow your heart
If you find it in your heart
My heartfelt sympathies on your loss
Wear one's heart on one's sleeve
He died of a broken heart
His heart is in the right place
Eat your heart out

She had a change of heart
Her heart was not in it
He has a heart of gold
A bleeding heart
A faint heart
Absence makes the heart grow fonder
With an aching heart
Open one's heart to
Pour one's heart out to
With a heavy heart

Introduction

Consider these famous quotes and common expressions for a moment. Collectively, they assume that the heart is not only a machine that pumps blood but also the seat of emotions (aching heart, broken heart), thought (reason, Pascal), and personality (bleeding heart, faint heart). Nobody says, "Follow your liver" or "Absence makes the kidneys grow fonder." Only a comedian would say, "The pancreas has its reasons of which reason knows nothing."

I cannot possibly address all the cultural symbolism that has collected around the heart. As we shall see, while these expressions and metaphors about the heart reflect centuries of folk wisdom, it seems that the ideas we acquaint in our collective unconscious with the heart as a center of thought, feeling, and personality are closer to recent discoveries in cardiac function than science previously assumed.

We shall explore the role the heart plays in our sense of self and our personality and show that sensitive recipients of heart transplants can experience changes in their own personality that dovetail with those of their donor. Such changes are further proof of information/memory being embedded in body cells and tissues as well as brain cells.

Heart Biology

First, some basics. The heart is a hollow muscular organ, located in the center of the chest. The heart and blood vessels constitute the cardiovascular (circulatory) system. The contraction of the muscle fibers in the heart is synchronized, like the gut, and highly controlled. Rhythmic electrical impulses (discharges) flow through the heart in a precise manner along distinct pathways and at a constant speed. The impulses originate in the heart's *natural pacemaker* (*the sinus* or *sinoatrial node*—a small mass of tissue in the wall of the *right atrium*), which generates a tiny electrical current. Some cardiac cells do the work of pumping blood (the muscle cells of *atria* and *ventricles, chambers of the heart*), while others exhibit electrical oscillations by which they regulate the rhythm of the heart.

The human heart contains an estimated two to three billion cardiac muscle cells, which account for about a third of the total number of cells in the heart. The balance includes smooth muscle and endothelial cells (cells that line the interior surface of blood vessels, lymphatic vessels, coronary arteries, and chambers of the heart). The heart also contains fibroblasts and other connective tissue cells, mast cells, and immune system-related cells. Recently, pluripotent cardiac stem cells have also been identified in the heart. These distinct cell groups are not isolated but constantly interact with each other.

The heart, like the gut, also contains an intrinsic nervous system that exhibits both short and long-term memory functions. According to J. A. Armour of Montréal's Hôpital du Sacré-Coeur the intrinsic nervous system of the heart consists of approximately forty thousand neurons called *sensory neurites* that relay information to the brain. It is possible that these neurons play a pivotal part in memory transfer.

The heart is equipped with its own unique conduction system, with regulatory influence from the central nervous system. Unlike the brain, conduction is not by neurons, but by special excitable muscle cells. Cardiac cells are *auto-rhythmic*; that is, they produce their own action potentials. This forms the basis of the heart's semiautonomous function. Cell-to-cell contact points called gap junctions connect cardiac cells—we have spoken of these before—that permit conduction of electrical signals by direct movement of ions from one cell to another. Gap junctions enable large areas of cardiac tissue to contract as a single unit.

The heart rate, or pulse, is the number of times the heart beats within a minute. The heart rate increases when the body needs more oxygen (such as during exercise) and decreases when the body needs less oxygen (as during rest). The rate at which the sinus node delivers its impulses (and thus governs the heart rate) is determined by two opposing parts of the autonomic nervous system—the sympathetic system speeds up the heart while the parasympathetic slows it down.

The sympathetic division works through a network of nerves called the sympathetic plexus and through the hormones epinephrine (adrenalin) and norepinephrine (noradrenaline), which are released by the adrenal glands and the nerve endings. The parasympathetic division works through a single nerve, the *vagus* nerve, which releases the neurotransmitter acetylcholine (FIG 8.1).

An intriguing study from the University of Buenos Aires has shown that when the heart is conditioned by a stimulus the same stimulus will subsequently activate it in a shorter time. The researchers remarked on the similarity between *heart memory* and the cortical processes of learning and retrieval in the brain. Along the same lines, researchers in Texas have shown that changes induced in cardiac activation rhythms persist long after the trigger that induced those changes is removed. Response to the same stimulus later on is much greater than the earlier response. The cardiologists concluded that the heart, like the nervous system, possesses the properties of **memory and adaptation**.

You will recall that in the chapter on bacteria we learned about how the gut microbiome affects the brain, both through the proteins it releases into the blood stream and by way of the vagus nerve. This is the same vagus nerve that supplies the heart and carries signals in both directions, from the brain to the heart and from the heart to the brain. When we encounter a dangerous situation, signals from the brain make sure that the heart beats faster. When we relax the heart slows down, which in turn affects the brain. There is heavy traffic along the vagus highway.

What is truly surprising and highly relevant to our investigation is the discovery that the heart also functions as an endocrine organ. In other words, just like the thyroid gland or the adrenal gland, it produces several hormones, including the *cardiac natriuretic peptide*.[1] This hormone exerts its effect on the blood vessels, on the kidneys, the adrenal glands, and on a large number of regulatory regions in the brain. It was also found that the heart contains a cell type known as *intrinsic cardiac adrenergic cells*. These cells release noradrenaline and dopamine neurotransmitters, once thought to be produced only by neurons in the CNS. More recently, it was discovered that the heart also secretes *oxytocin*, commonly referred to as the love or bonding hormone. In addition to its functions in childbirth and lactation, and the establishment of enduring pair bonds, recent evidence indicates that this hormone motivates and enables humans to like and empathize with others, comply with group norms and cultural practices, and extend and reciprocate trust and cooperation. Concentrations of oxytocin in the heart were found to be as high as those in the brain.

The heart generates the body's most powerful and most extensive rhythmic electromagnetic field. Compared to the electromagnetic field produced by the brain, the electrical component of the heart's field is about sixty times greater in amplitude, and reaches every cell in the body. The magnetic component is approximately five thousand times stronger than the brain's magnetic field and can be detected several feet away from the body with sensitive magnetometers.

The rhythmic beating patterns of the heart change significantly as we experience different emotions. Negative emotions, such as anger or frustration, are associated with an erratic, disordered, incoherent pattern in the heart's rhythms. In contrast, positive emotions, such as love or appreciation, are associated with a smooth, ordered, coherent pattern. In turn, these changes in the heart's cadence create corresponding changes in the structure of the electromagnetic field generated by the heart, measurable by a technique called spectral analysis.

The brain, the heart, and all other bodily organs produce electromagnetic waves that people close by, depending on how sensitive and open they are to such stimuli, may resonate to in totally unconscious ways. These noncognitive responses we call "gut feelings" or "intuition." It is also what happens when a mother gazes directly and lovingly into her infant's eyes and the child returns her smile attentively. In light of what we are learning here, it is safe to assume that this communication is by electromagnetic waves between the mother's brain and her child's brain and leads to bonding and attachment.

This point is well illustrated by a study of twenty-two heterosexual couples, age twenty-three to thirty-two, who had been together for at least one year. The couples were subjected to several two-minute scenarios as electroencephalography (EEG) measured their brainwave activity. The scenarios included sitting together not touching, sitting together holding hands, and sitting in separate rooms. Then they repeated the scenarios as the woman in each pair was subjected to mild heat pain on her arm. Merely being in each other's presence, with or without touch, was associated with some brain wave synchronicity. Brain wave synchronicity increased the most when the couple held hands while the woman was in pain. I wish someone would repeat this experiment using electrocardiograms (ECGs). I would wager that the ECGs would also show synchronicity.

Of course, you do not have to be in a loving relationship to feel good or to feel supported by another person. Several years ago, while on a lecture tour in Spain, I became acutely ill and needed to go to the emergency department of a local hospital. Since I speak no Spanish, my translator, a very warm and caring person, came with me. He was allowed to accompany me into the surgery, where I underwent a painful procedure—during which I reached out to him. He held my hand and honestly, I felt so very grateful because it helped me tolerate the pain so much better.

Beyond pair bonding there is growing research into the physiology of group bonding. The ability to coordinate our actions with those of others is crucial for our success as individuals and as a species. Therefore, it is not surprising that collective activities such as working together or singing or dancing, to name a few, are accompanied by cardiac and respiratory pattern synchronization. I am reminded here of a study from the Max Planck Institute in Berlin on eleven singers and one conductor engaged in choir singing. They observed that their heart rates and breathing accelerated and decelerated simultaneously during singing. The researchers hypothesized that their results showed that oscillatory coupling of cardiac and respiratory patterns provide a physiological basis for interpersonal action coordination.

From Bar-Ilan University in Israel we learn that hearts that drum together, in fact, beat together. Analysis of the data from their study demonstrated that group drumming elicited an emergence of cardiac synchronization beyond what could be expected randomly. Further, behavioral synchronization and enhanced physiological synchronization while drumming each uniquely predicted a heightened experience of group cohesion. Finally, the researchers showed that higher physiological synchrony also predicts enhanced group performance at a later time in a different group task.

How is this relevant? These experiments have shown that when people are in emotional synchrony, so are their brains and hearts, and so is their breathing. Both the brain and the heart generate extensive electromagnetic fields with the heart having the much more powerful one. When two or more people are in close proximity these electromagnetic vibrations will be picked up and responded to by their respective organs. It is like two people standing six feet apart, each of them holding a tuning fork. When one person strikes their tuning fork, the other's tuning fork starts to vibrate in the same rhythm. When our hearts are on the same wavelength with the heart of another person, and to a lesser extent, when our brains synchronize, mutual attraction takes place and we feel emotionally close to that person.

Remember the kidnapping-and-cross-fostering study in the chapter on genetics? European honeybees raised among more aggressive African bees not only became as belligerent as their new hive mates, they came to genetically resemble them. And vice versa. Proof that in a very short time **the social environment can change gene expression and behavior.** In like manner, when two or more people spend time together their organs synchronize, particularly the heart, which then fosters social bonding.

Heart and Personality

A recent study from Denmark's Aarhus University Hospital takes us in a different direction. Their study involved 10,632 adults born between 1890 and 1982. Using medical registries and medical records from all Danish hospitals they identified adults who were diagnosed between 1963 and 2012 with congenital heart disease. These individuals had a 60 percent higher rate of dementia compared to the general population. The risk was 160 percent higher in people with congenital heart disease in people younger than sixty-five. Now, why should that be?

Previous studies have linked negative emotions, including depression, anxiety, and anger, to a heightened risk of heart disease. Because these emotions tend to overlap and coexist, it's been difficult to assign a relative importance to any one of them.

When it comes to myocardial infarction, the medical term for a heart attack, evidence continues to accumulate that if a person develops a major depression following a heart attack, a rather common occurrence, they will be consistently at a threefold increased risk of death. What all of this shows is the extent to which the heart affects the brain and the rest of the body and vice versa.

According to new research from Rice University and Northwestern University, people who recently lost a spouse are more likely to have sleep disturbances that make them more vulnerable to develop inflammations, which in turn raises their risk to develop cardiovascular illness and death. This is just another example of the perfectly designed ecosystem our body represents. The moment one of its elements is changed everything else is affected.

Let's move from depression to schizophrenia for a moment. The leading cause of death among people suffering from schizophrenia is coronary artery disease. The average life expectancy of the general population in the United States is seventy-six years (seventy-two years in men, eighty years in women), compared to sixty-one years (fifty-seven years in men, sixty-five years in women) among patients with schizophrenia. Thus, individuals with schizophrenia have approximately a 20 percent reduced life expectancy relative to the general population.

At the 2018 Congress of the European Psychiatric Association a report on a nationwide study of thirty-seven thousand schizophrenia patients in Denmark revealed that mortality rates were highest in patients with schizophrenia who had experienced an acute myocardial infarction.

Cardiologists believe that factors such as antipsychotic medications, cigarette smoking, obesity leading to high cholesterol, hypertension, insulin resistance, and diabetes are responsible for escalating morbidity and mortality in schizophrenic patients. In addition, patients with schizophrenia are thought to have less access to medical care and when they do are less compliant with their doctors' advice and less likely to take prescribed medications.

I think more to the point is the effect of severe acute stress or prolonged chronic stress that can alter biological systems in a way that, over time, adds up to "wear and tear" and eventually leads to illnesses such as heart disease, stroke, and diabetes. In addition, stress causes increased cardiac sympathetic activity and decreased parasympathetic activity, making the myocardium prone to arrhythmias, and likely responsible for some of the cardiac changes seen in these patients.

Sustained anger and anxiety are other factors that can disrupt cardiac function by changing the heart's electrical system, hastening atherosclerosis, and increasing systemic inflammation. According to Laura Kubzansky of the Harvard School of Public Health, much depends on the balance between negative and positive emotions in a person.

In a 2007 study, Kubzansky followed more than six thousand men and women aged twenty-five to seventy-four for twenty years. She found that emotional vitality—a sense of enthusiasm, of hopefulness, of engagement in life, and the ability to face life's stresses with emotional balance appeared to reduce the risk of coronary heart disease. The protective effect was distinct and measurable, even when taking into account such wholesome behaviors as not smoking and regular exercise. Kubzansky has found that **optimism cuts the risk of coronary heart disease by half.**

Research in the relatively new discipline of neurocardiology has confirmed that the heart is a sensory organ and acts as a sophisticated information encoding and processing center that enables it to learn, remember, and make independent functional decisions that do not involve the cerebral cortex. Additionally, numerous studies have demonstrated that patterns of cardiac signals to the brain affect autonomic regulatory centers and higher brain centers involved in perception and emotional processing.

Heart Transplants

In heart transplant surgery, the heart's vascular and neuronal connections from its donor are cut. Then this heart is transferred into the chest of the recipient, where it will take months to connect with the nerves that formerly supplied that person's heart. During this time the transplanted heart is able to function in its new host only because it possesses its own intrinsic nervous system that operates and processes information independently of the brain. Please keep this in mind as we forge ahead.

Anecdotal though they may be, there are numerous accounts in popular books and magazines of heart transplant recipients experiencing distinct changes in personality and behavior following transplant surgery. When we consider the possibility of such personality changes—what in effect we are investigating is **memory transfer**. Remember the studies by David Glanzman, who successfully transferred a memory from one marine snail to another, by extracting from the nervous systems of those sensitized or trained (i.e., exposed to electrical shock) snail donors and injecting them into the body of untrained snail recipients.

In a similar vein, Shelley L. Berger at the University of Pennsylvania, experimenting with mice, discovered that after a conditioning and learning trial a *metabolic enzyme, acetyl-CoA synthetase 2,* affected epigenetically key memory genes within the nucleus of neurons. In other words, RNA and an enzyme in the nucleus of neurons is likely involved in memory storage.

And of course, there is Shomrat and Levin's research on planaria that led them to conclude **that traces of memory of the learned behavior are retained outside the brain**. They suggested that this was by way of mechanisms that include the cytoskeleton, metabolic signaling circuits, and the gene regulatory networks.

And, of course, all the other research we considered that showed that memories are distributed throughout the body, not just the brain.

What is the relationship among cellular memory, organ transplantation, and corresponding personality transfer? There is a paucity of solid scientific studies in the medical literature on personality changes following heart transplants. No doubt, the prevailing belief that memories are stored in the brain and nowhere else discourages transplant recipients from even entertaining the idea that their personality may be affected by that of the organ donor or to speak of it with their families, friends, or physicians. Similarly, researchers are held back from studying this subject for fear of being labeled by their peers as unscientific, anecdotal or, worse of all, flaky. In this work, I ask you to examine the evidence and not prejudge the issue.

A retrospective inquiry on heart transplant patients at University Hospital in Vienna found three patients out of a total of forty-seven reporting a distinct change of personality due to their new hearts. Interestingly, a single, forty-five-year-old patient who received a heart from a seventeen-year-old boy said that following heart transplantation he started to enjoy loud music and dreamed of having a car with a good stereo, something he did not desire prior to the surgery. The authors of the paper were, true to form, rather dismissive of such reports, writing, "Verbatim statements of these heart transplant recipients show that there seem to be severe problems regarding graft incorporation, which are based on the age-old idea of the heart as a center that houses feelings and forms the personality."

A study of heart transplants in Israel was more positive. In their study many recipients endorsed "fantasies and displayed magical thinking."

Specifically, 46 percent of the recipients had fantasies about the donor's physical vigor and powers, 40 percent expressed some guilt regarding the death of the donor, 34 percent entertained an overt or covert notion of having acquired some of the donor's personality characteristics along with the heart. In other words, one in three recipients felt that their personality adopted some of the characteristics of the donor's personality.

Mitchell and Maya Liester at the University of Colorado, in a 2019 paper, after reviewing the literature on heart transplants, hypothesized that the acquisition of donor personality characteristics by recipients following heart transplantation may occur by way **of transfer of cellular memory.** They suggested that cellular memory consisted of epigenetic memory, DNA memory, RNA associated memory, horizontal gene transfer, protein memory, and memory stored in the electromagnetic field of the heart. The authors concluded that heart transplant recipients provide evidence that personality is not limited to the brain but may also be stored in the heart, some of which may be transferred via heart transplantation from donor to recipient.

About ten years ago I read a paper by Paul Pearsall and fellow researchers in which they discussed ten cases of heart or heart-lung transplants. Pearsall had interviewed transplant recipients, their families, and the donor's family, while his associates, Gary Schwartz and Linda Russek, examined parallels between the donor and recipient. According to the authors, the recipients experienced profound changes in their lifestyles: "Changes in food, music, art, sexual, recreational, and career preferences, as well as specific instances of perceptions of names and sensory experiences related to the donors."

Here is one of their case studies.

Case 5

The donor was a 19-year-old woman killed in an automobile accident. The recipient was a 29-year-old woman.

The donor's mother reported:

"My Sara was the most loving girl. She owned and operated her own health food restaurant and scolded me constantly about not being a vegetarian. She was a great kid. Wild, but great. She was into the free love thing and had a different man in her life every few months. She was man-crazy when she was a little girl and it never stopped.

She was able to write some notes to me when she was dying. She was so out of it, but she kept saying how she could feel the impact of the car hitting her. She said she could feel it going through her body."

The recipient reported:

"You can tell people about this if you want to, but it will make you sound crazy. When I got my new heart, two things happened to me. First, almost every night, and still sometimes now, I actually feel the accident my donor had. I can feel the impact in my chest. It slams into me, but my doctor said everything looks fine.

I couldn't tell him, but what really bothers me is that I'm engaged to be married now. He's a great guy and we love each other. The sex is terrific. The problem is, I'm gay. At least, I thought I was. After my transplant, I'm not . . . I don't think, anyway . . . I'm sort of semi- or confused gay. Women still seem attractive to me, but my boyfriend turns me on; women don't. I have absolutely no desire to be with a woman. I think I got a gender transplant.

Also, I hate meat now. I can't stand it. I was McDonald's biggest moneymaker, and now meat makes me throw up. Actually, when I even smell it, my heart starts to race. But that's not the big deal. My doctor said that's just due to my medicines."

The recipient's brother reported:

"Susie's straight now. I mean it seriously. She was gay and now her new heart made her straight. She threw out all her books and stuff about gay politics and never talks about it anymore. She was really militant about it before. She holds hands and cuddles with Steven just like my girlfriend does with me. She talks girl-talk with my girlfriend, where before she would be lecturing about the evils of sexist men.

And my sister, the queen of the Big Mac, hates meat. She won't even have it in the house." [Permission for these quotes granted by *Nexus* magazine.]

⁂

Schwartz believed that memory exists in every heart. It is only a question of how many people who receive a heart will become aware of that information, or be significantly influenced by it.

There are many descriptions in the popular press of heart transplant recipients taking on the personality traits of their donor. Perhaps the best known account is that of Claire Sylvia, a former professional dancer. Sylvia received a heart from an eighteen-year-old boy who died in a motorcycle accident. After the surgery she started craving beer and KFC fried chicken, stuff she never liked before. "My daughter said I even walked like a man." Wanting to understand the changes she was experiencing, she sought out the family of her donor and learned that these foods were his favorites.

As we read these reports, we should keep in mind that the prevailing belief that memories are stored in the brain and nowhere else, discourage transplant recipients from readily accepting changes in their personality that parallel that of the organ donor or, to speak of these with their families, friends, or physicians. Fear of being labeled as weird or crazy no doubt dampens many a heart recipient's enthusiasm to share their experiences with others.

Records of memory transfer following kidney transplants, lung transplants, or skin grafts are virtually nonexistent. Also, there are no records of humans undergoing any personality changes following transplantation of porcine or bovine heart valves or other tissues or organs from animals. Reported cases of memory transfer seem to be specific to heart transplants, which makes sense considering the very special organ the heart is.

Generally speaking, scientists distrust and readily dismiss as mystical or "New Age" patients' self-reports about experiences that differ from what they were taught in college. Mainstream cardiologists dismiss the idea that an organ such as the heart can impart personality characteristics of the donor to the recipient. John Schroeder of Stanford Health Care speaks for this orientation when he says, "The idea that transplanting organs transfers the coding of life experiences is unimaginable." Because Schroeder cannot imagine it, the phenomenon cannot exist. Not exactly a scientific refutation.

Another skeptic, Heather Ross, who was medical director of the heart transplant program at the Peter Munk Cardiac Centre in Toronto, scoffed, "It's been out in the pop culture for a long time, for sure. There is no scientific evidence for such a thing."

When personality changes have been observed following transplants, the kinds of explanations entertained by skeptics include effects of the immunosuppressant drugs, stress of the surgery, psychosocial stress, pre-existing psychopathology, and statistical coincidence. To my mind, these explanations are insufficient to explain the findings.

Let's be realistic. Scientists, like the rest of us, are not always paragons of rationality, objectivity, and open-mindedness. Especially when they attack other scientists' ideas divergent from their own. You need to take what these dyed-in-the-wool apologists for the status quo say not with a grain of salt but a whole ton of salt.

The Good News

In a 6,626-participant study, researchers at the University of Bordeaux found that cardiovascular health in older people is associated with lower risk of dementia and lower rates of cognitive decline.

A new scientific advisory reaffirms the American Heart Association's recommendation to eat fish—especially those rich in omega-3 fatty acids—twice a week to help reduce the risk of heart failure, coronary heart disease, cardiac arrest, and the most common type of stroke (ischemic).

Training in techniques to increase group coherence and heart rhythm synchronization will correlate with improved communication and boost behavior intended to promote social acceptance and friendship.

Numerous studies of various populations, regardless of geography or culture, have found that individuals who have close and meaningful relationships benefit by reducing the risk of mortality and morbidity, have improved outcomes in pregnancy and childbirth, and live happier, healthier, and longer lives.

Summary

There is no doubt that the signals the heart sends to the brain influence the function of higher brain centers involved in perception, cognition, and emotional processing, and vice versa. Because of this feedback loop, it is not surprising that people who suffer from depression or schizophrenia or simply too much stress have been shown to be at a higher risk to develop heart disease than emotionally healthy individuals. The heart is very much an essential part of one integral, interconnected, multilevel body-mind system.

The heart's brain is an intricate network of several types of neurons, neurotransmitters, proteins, and immune cells similar to those found in the enskulled brain. This *heart brain* enables the heart to act independently of the cranial brain—to learn, remember, and even feel and sense.

Like the gut but even more so, the heart is connected to the brain by the vagus nerve and the autonomic nervous system. The heart is also part of the body's endocrine system and secretes peptides (hormones) of its own. Furthermore, it generates a unique electromagnetic field that affects the rest of the body and extends several feet beyond.

How do we explain instances of immediate, lightning attraction, "love at first sight," repulsion, or distrust of an individual we have just met? Such occurrences are often spoken of or dismissed out of hand as based on "intuition." I suggest that the heart's electromagnetic field operates as an "energetic" communication system below our conscious awareness and is responsible for these phenomena. The heart seems to be intimately involved in the generation of psychophysiological coherence.

Heart transplant surgery is not simply a question of replacing a diseased organ with a healthy one. It is more than that, as reports in the literature of heart transplant recipients confirm their experiencing changes in their own personality that dovetail with those of their donor. Such personality changes, at least in some people, would be expected in light of our finding that information/memory is stored in body cells and tissues. Cardiac cells would be no exception to that, and if anything, more likely to carry personal data.

Restoring memories to people who have suffered strokes with accompanying amnesia or who have been diagnosed with Alzheimer's could one day be accomplished by successfully identifying and isolating specific cardiac cells carrying memories. This would represent a life-changing gift to millions of people. We know that electric stimulation of certain parts of the brain in an awake subject unearths long "forgotten" memories or musical melodies or smells. So, the idea of looking for such buried recollections in cells is not so far-fetched. Of course, it may take scientists a while to discover the means of achieving this.

KEY TAKEAWAYS

- **The heart is not just a pump; kidneys don't just purify our blood; the lungs are more than breathing machines.**
- **The heart, like the nervous system, possesses the properties of memory and adaptation.**
- **The heart acts as a synchronizing force within the body, a key carrier of emotional information as well as other information related to personal identity.**
- **Sensitive transplant patients may evidence personality changes that parallel the experiences, likes, dislikes, and temperament of their donors.**
- **It is comforting for the donors' families to see evidence of not just their loved ones giving someone a chance at life, but to know that a bit of them lives on within the recipient.**

- Such changes are further proof of information/memory being embedded in somatic cells and tissues as well as brain cells.

- It seems that the ideas we carry in our collective unconscious about the heart as a center of thought, feeling, and personality are closer to modern science than science previously postulated.

CHAPTER NINE

THE BRAIN-MIND CONUNDRUM AND THE RISE OF QUANTUM BIOLOGY

The universe is not only stranger than we think, but stranger than we can think.

—Werner Heisenberg, 1932 Nobel Prize winner, "for the creation of quantum mechanics"

Introduction

In this chapter I want to address central issues of consciousness, free will, and the brain-mind relationship. The modern scientific worldview is predominantly grounded in classical Newtonian physics. It considers matter the only reality. This scientific view is referred to as *materialism*. A related assumption is the notion that complex things can be understood by reducing them to the interactions of their parts, usually smaller, simpler, or more fundamental bits such as atoms and electrons. This is called *reductionism*. Materialism and reductionism are science's Tweedledum and Tweedledee.

The research cited on the preceding pages flows from this well of classical science. Based on this science it would be reasonable for the reader to conclude that our genetic endowment combined with epigenetic changes in response to environmental challenges has produced the mind, consciousness, the choices we make, even our beliefs, likes, and dislikes. Biological phenomena such as electrical charges, neurotransmitters, and hormones are considered responsible for and fully explain these processes. Many neurologists, philosophers, and psychologists are of the opinion that if all of the myriad factors that contributed to the construction of our bodies (including the brain) were known we could precisely predict how a person would act in any situation at any time. In other words, we have been programed like a computer. The mind arises from the operations of the brain. Free will is an illusion.

Similarly, using the language of neurons and cortical excitation, these scientists hold that the brain generates consciousness. When the brain suffers injury, consciousness deteriorates; when a person dies, the brain dies, and consciousness ceases.

Scientists armed with ever more precise fMRIs, EEGs, and the other tools of the materialist and determinist position, have mapped our brains successfully and located areas responsible for sight, hearing, executive functions, and many others. They have been very successful in identifying the brain's neural circuitry, electrical conduction along axons, or chemical diffusion across synapses. But the terra firma of materialism becomes far less firm, far shakier when neuroscientists attempt to understand with the devices and approaches of classical science the more profound mystery of the mind. How does a three-pound organ with the consistency of Jell-O create feelings of awe or empathy?

Scientists have shown that information reaching the brain is broken down into separate processing streams. But no one has yet found any "place" where all the information is assembled into a complete picture of what is being felt, thought, or experienced. These scientists are not even close to discovering how the brain creates conscious experience. Will, reason, or the mind do not currently have any precisely identified neural correlates.

How can the firing of billions of neurons give rise to thoughts, imagination, art, or appreciation of beauty or complex feelings like love, hate, or happiness? Somehow, brain processes acquire a subjective aspect, which at present seem impenetrable to classical science. Enter post-materialism science and quantum biology.

Consciousness

Since the beginning of time, men and women have tried to come to grips with the three fundamental mysteries of life: consciousness, free will, and God—without much success. Are these "hard" problems, as philosopher David Chalmers characterized consciousness, or are they truly insoluble "mysterian" problems, as philosopher Owen Flanagan called them?

Consciousness is generally understood to mean that an individual not only has an idea, recollection, or perception but also *knows* that he or she has it. This knowledge accords individuals a sense of self and agency. It encompasses both the experience of the outer world ("it's sunny") and one's inner world ("I'm happy"). Consciousness, as I see it, is a faculty of the mind.

Essential to generating a sense of self is our ability to remember, which is why this work has focused so much attention on elucidating where memories are stored in the body and how they are passed from one generation to the next.

Most biologists believe that in the process of evolution, consciousness arose when cortical neurons in living organisms reached an inflection point denoting a certain degree of complexity. This theory implies that consciousness emerged, just like life itself, from inanimate matter. Therefore, old-school neuroscientists have ascribed to the brain a singular and dominant importance, a fact that has for a long time discouraged research into differences between brain and mind, as well as the origins of consciousness and free will.

The study of consciousness is complicated because of the inherent difficulty of applying the conscious mind to study itself. According to Ezequiel

Morsella of San Francisco State University, "Consciousness is the middleman, and it doesn't do as much work as you think." Morsella, in contrast to myself (as I shall explain), proposes that the conscious mind is like an interpreter helping speakers of different languages communicate, neither the source of the original communication nor capable of acting upon it.

The presently accepted view among neuroscientists is that consciousness occurs after-the-fact, as an *epiphenomenon* (the same *epi* as in epigenetics), a function of the brain. Actually, Pierre Jean Georges Cabanis (1757–1808), a French physiologist, stated this position already two hundred years ago as: "The brain secretes thought like the liver secretes bile." However, it was not until 1991, when mainstream cognitive science and philosophy fully adopted American philosopher Dan Dennett's term *epiphenomenalism* as their party line.

In the past, consciousness was thought to reside in the prefrontal cortex. More recently, the prefrontal cortex is seen as regulating levels of consciousness by way of reciprocal interactions with subcortical arousal systems as well as neural circuits of attention, working memory, and verbal and motor processes.

Some authors, based on lesion data, electrical or magnetic stimulation data, and functional brain imaging data, argue that both content-specific and full neural correlates of consciousness are mainly located in the posterior part of the brain, encompassing the parietal, occipital, and lateral temporal lobes.

In 1992 Sir John Carew Eccles, an Australian neurophysiologist and philosopher who won the 1963 Nobel Prize in physiology and medicine for his work on the synapse, proposed that consciousness likely occurs in dendrites. Two decades later, Karl Pribram, the eminent brain scientist, psychologist, and philosopher, called by his colleagues the "Magellan of the Mind," added his support to this theory. Obviously, this is a work in progress.

A prominent international group of cognitive neuroscientists, neuroanatomists, neuropharmacologists, neurophysiologists, and computational neuroscientists gathered at the University of Cambridge in 2012 to reassess the neurobiological substrates of conscious experience and related behaviors in human and nonhuman animals. At the conclusion of their meeting, they published *The Cambridge Declaration on Consciousness,* according to which, "The absence of a neocortex does not appear to preclude an organism from experiencing affective states. **Convergent evidence indicates that nonhuman animals including all mammals and birds, and many other creatures, including octopuses, have the neuroanatomical, neurochemical, and neurophysiological substrates of conscious states along with the capacity to exhibit intentional behaviors."**

In this context, it is not surprising that anesthetists as a group stand out from among other scientists and philosophers as having contributed much pioneering research on consciousness. Ether's ability to induce loss of consciousness was first demonstrated on a tumor patient at Massachusetts General Hospital in Boston in 1846, within a surgical theater that later became known as "the Ether Dome." So consequential was the procedure that it was captured in a famous painting, *First Operation Under Ether*, by Robert C. Hinckley. The Ether Dome is still in daily use. During my studies at Massachusetts General, I was asked to present a paper in this famous hall. To do so was both thrilling and terrifying.

When people are administered an anesthetic, they seem to lose consciousness or at least they stop reacting to their environment. Anesthetic agents do not suppress brain function globally but exert dose-dependent effects on specific brain systems that block the perception of pain.

The question that is central to consciousness under anesthesia is whether consciousness is fully lost during anesthesia or whether it persists but in an altered state? A joint research project, "The Conscious Mind: Integrating Subjective Phenomenology with Objective Measurements," from Finland's University of Turku has explored in depth this question. Their studies revealed that the brain processes sounds and words even though the subject does not recall it afterward. The findings indicate that the state of consciousness induced by anesthetics is similar to natural sleep. While sleeping, people dream and the brain observes the occurrences and stimuli in the environment subconsciously. Under anesthesia, as in sleep, we can be of two minds. Classical neuroscience is unable to provide an explanation of this phenomenon.

The Volitional Self

Each of us experiences ourselves as being a continuous and distinctive person over time, built from a rich set of autobiographical memories. Of the many unique experiences within our inner universe, one is the experience of *being me*, in academic terms, the sense of *selfhood*. Conscious selfhood in classical psychology is understood as a complex construct generated by different parts of the brain communicating with each other.

The *volitional* self involves experiences of intention and of agency—of urges to do this or that, and of self-control. The relationship between causality and intentionally willed action remains obscure. Free will is the part of our minds that has caused the most controversy over the ages among philosophers, religious scholars, scientists, and writers (think Dostoevsky or Camus).

Our society assumes that human beings possess volitional selves and, consequently, are capable of moral reasoning. In this context, free will allows people to choose between good and evil, and the law punishes antisocial behavior. Free decision-making is a cornerstone of our society supported by all mainstream faiths and jurisprudence but not by old-school deterministic neuroscience.

An example of the research supporting that orientation is the work pioneered by Benjamin Libet. Libet found that when study participants were asked to perform a specific task, their brain activity preceded their actions. Later studies, using various techniques, claimed to have replicated this basic finding.

Veljko Dubljević, an assistant professor of philosophy at North Carolina State University who specializes in research on the neuroscience of ethics and the ethics of neuroscience and technology, reviewed forty-eight studies, ranging from Libet's landmark 1983 paper through 2014. Basically, he found that interpretation of these studies was strongly influenced by the subjective beliefs of the authors rather than by a careful and objective analysis of the results themselves. Dubljević also observed that a significant subset of studies that assessed the location in the brain where activity was taking place was not related to will or intent to complete a task. While the Libet approach may be useful for examining how stimuli affect temporal judgments, the link between this and free will or moral responsibility is anything but clear. Generally speaking, the literature is rather contradictory and does not reveal a consistent picture of the functional neuroanatomy of volitional behavior.

In *The Emperor's New Mind,* Sir Roger Penrose—an esteemed figure in mathematical physics at Oxford University and recipient of the 2020 Nobel Prize in physics for the discovery of black hole formation—is a robust prediction of the general theory of relativity, suggested that *quantum mechanism* (more on this later) explains Libet's backward time effects.

Mind Over Matter

Through the 1990s, Dr. Masaru Emoto, a Japanese author, researcher, photographer, and entrepreneur, performed a series of experiments observing the physical effect of words, prayers, music, and environment on the crystalline structure of water. Emoto exposed water to different variables and subsequently froze it so that crystalline structures formed.

In one series of experiments, Emoto taped different words, both positive and negative, on containers filled with water. The water container stamped with positive words produced more symmetrical and aesthetically pleasing crystals than the water in containers stamped with dark, negative phrases. "Water is the mirror of the mind," according to Emoto. Emoto's research demonstrates that human vibrational energy; words spoken or written, feelings, and music affect the molecular structure of water. But it is not only water that our thoughts and feelings affect. Recall from the previous chapter how the rhythmic, even patterns of the heart become erratic, disordered, and incoherent with negative emotions, such as anger or frustration. In contrast, positive emotions, such as love or appreciation, are associated with a smooth, ordered, coherent pattern.

Many people have replicated Emoto's rice experiment, including two of my friends.[1] You can see the results of their experiments below (FIG 9.1a, 9.1b).

They started their experiment on February 1, 2007. They kept two labeled jars of cooked white rice on top of their piano, not too far apart so that they would have the same light and room temperature, etc. Toward the rice on the left, they daily directed their voices, saying, "Thank you! You're beautiful!" Toward the rice on the right, they said, "You fool! You stink!"

You can see what the jars of rice on their piano looked like three months and six months later in figures 1a and 1b in the image insert.

What a difference. The rice on the right is almost totally rotten while the one on the left is still pretty healthy. If rice reacts this way to emotionally charged words, how do you think people's bodies react?

Water makes up 80 percent of rice as well as of our bodies. While there is much psychological evidence for the power of words, there is a dearth of biological research on this subject. An example of the former approach is the research of David Chamberlain, San Diego psychologist and one of the early pioneers of pre- and perinatal psychology. According to Chamberlain, birth memories that arise in the course of insight-oriented psychotherapy illustrate how babies can be stung and poisoned for decades by unkind remarks such as "What's wrong with her head?" or "Wow, this looks like a sickly one."

Several years ago, I visited a large neonatal intensive care unit (NICU) in a teaching hospital. There were thirty-six isolettes, or incubators, with babies in the room (FIG 9.2). About half of them had their names displayed on the side of their incubators. The other half had no names. I asked the charge nurse why these babies were without names. She said, "Because their parents do not want to get too attached to them, in case they die." I wish someone had done research comparing the health, physical and psychological, of these two groups of children over a good length of time. I have absolutely no doubt that the children whose parents addressed them by their given names would be healthier and live longer compared to the nameless group.

Another example of how words can affect the mind and how the mind in turn can affect the body is hypnosis. Hypnosis is best described as an altered state of consciousness, similar to relaxation, meditation, or sleep. Many people think that it was Sigmund Freud who originated depth psychology in his 1899 book, *The Interpretation of Dreams*. However, depth psychology and psychoanalysis actually predated Freud and had their roots in the practices of mesmerism, hypnosis, and earlier esoteric disciplines. Modern hypnosis was first popularized by the Austrian medical doctor Franz Anton Mesmer (1734–1815), who described its specific states and called it "animal magnetism." James Braid (1795–1860), introduced the term *hypnosis* from the Greek *hypnos*, meaning "sleep," as he considered hypnosis to be a "nervous sleep." Traditionally, psychologists and neuroscientists have been skeptical of hypnosis and distrustful of participants' subjective reports of profound changes in perception following specific suggestions. However, the advent of cognitive neuroscience and the application of neuroimaging methods to hypnosis has brought about the validation of participants' subjective responses to hypnosis. Therefore, it is not surprising that in 1958 the American Medical Association suggested that hypnosis should be included in the curriculum of medical schools, and in 1960 the American Psychological Association officially acknowledged the therapeutic use of hypnosis by psychologists.

Individuals suffering from chronic pain, irritable bowel syndrome, and PTSD have benefited from hypnosis. Psychotherapists have also successfully used hypnosis in the service of age regression and the uncovering of past traumas. What hypnosis teaches us is that the words of a person who is perceived by the subject as having certain powers can change bodily movement ("Your arm feels light as a feather, let it rise toward the ceiling") or provide analgesia for painful procedures or help the patient unearth long-forgotten memories.

As we shall see, hypnosis and placebo have much in common. While hypnosis-like phenomena have a documented history going back thousands of years, accounts of placebo effects span only several centuries. With the rise of biological psychiatry and the "pharmacological revolution," drug trials have taken a central place in clinical research. These clinical trials increasingly incorporate placebo-controlled conditions as part of their paradigms. So, what is a placebo?

A placebo is an inactive treatment, like a "sugar pill." In fact, a placebo may be a pill, tablet, injection, medical device, or suggestion. Whatever the form, placebos are meant to look like the real medical treatment that is being studied except they do not contain the active drug. Using placebos in clinical trials helps scientists better understand whether a new medical treatment is safer and more effective than no treatment at all. This is not always easy because some patients improve in a clinical trial even when they don't receive any active medical treatment or substance during the study. This is called the *placebo effect.* The placebo effect describes any psychological or physical effect that a placebo has on an individual.

Placebos have been shown to produce measurable physiological changes, such as an increase in heart rate or blood pressure. Placebos can reduce the symptoms of numerous conditions, including Parkinson's disease, depression, anxiety, irritable bowel syndrome, and chronic pain. Researchers have repeatedly shown interventions such as "sham" acupuncture to be as effective as acupuncture. Sham acupuncture uses retractable needles that do not pierce the skin.

Placebo interventions vary in strength depending on many factors. For instance, an injection causes a stronger placebo effect than a tablet. Two tablets work better than one, capsules are stronger than tablets, and larger pills produce greater reactions. One review of multiple studies found that even the color of pills made a difference to the placebo results. The positive health benefit that a patient experiences in response to a placebo is a function of the symbols, rituals, and behaviors embedded in their clinical encounter. Part of the power of the placebo lies in the **expectations** of the individual receiving them. These expectations can relate to the treatment, the substance, or the prescribing doctor. If these expectations are positive, the patient will have a positive response to the placebo and vice versa. This is very similar to hypnosis. The more the subject expects the hypnosis to work, the deeper they will enter a hypnotic trance. In essence, a person's mind-set, whether it is under hypnosis or on the receiving end of a placebo "drug," determines the results. A person expecting a certain outcome, such as pain relief, will by way of his mental operations initiate a cascade of physiological responses (hormonal, immunological, etc.) that will cause effects similar to what a medication might achieve. Like hypnosis, placebos demonstrate clearly the power the mind has over matter.

This process of expectations extends right into our brains. It has been by now well established that we humans depend on our senses to perceive the world, ourselves, and each other. People rarely question how faithfully our sense organs interpret external physical reality. During the last twenty years, neuroscience research has revealed that the cerebral cortex constantly generates predictions on what will happen next, and that neurons in charge of sensory processing only encode the difference between our predictions and the actual reality.

A team of neuroscientists of TU Dresden headed by Katharina von Kriegstein presents new findings that show that not only the cerebral cortex, but the entire auditory pathway, represents sounds according to prior **expectations**. The Dresden group has found evidence that this process also dominates the most primitive and evolutionarily conserved parts of the brain. **What we perceive is deeply influenced by our subjective beliefs.**

And where are expectations and predictions located? In the absence of a neurological substrate for such, we must assume it is the mind.

Drilling deeper into the operations of the mind we find reports in the literature of out-of-body experiences (OBEs) and near-dath experiences (NDEs) most instructive. Philosopher and psychiatrist Raymond Moody, whose 1975 book, *Life After Life*, detailed the experiences of more than one hundred people who survived "clinical" death but were revived, coined the term *near-death experiences* (NDEs). The book sold more than thirteen million copies, which if nothing else, proves the general public's fascination with this subject.

NDEs are triggered during singular life-threatening episodes when the body is injured by blunt trauma, a heart attack, asphyxia, or shock. When revived, these people tell us what it was like for them to exist suspended between life and death. Such people share strikingly similar narratives: they speak of having experienced wonderful sensations of peaceful tranquility and happiness, seeing a welcoming golden glow, often at the end of a tunnel, being greeted by deceased relatives, or detaching from their body and floating above it and even flying off into space. They often feel obliged to make a choice: remain in this other world and exit life as they knew it, or return to life.

Usually, they relate perceiving themselves from a perspective above or to the side of their physical body and describe accurately the conversations of the medical staff present in the room as well as the medical interventions that were performed on them. They claim that their spirit left their body and observed what they shouldn't have been able to observe given that they were clinically dead and lying prone on a bed. The majority of NDE survivors become wholly transformed by the experience. They grow into more caring, less self-centered persons, more engaged with helping others and less afraid of dying.

The phenomenon is remarkably consistent across cultures and religions, and has been reported even by children and toddlers. This is noteworthy because skeptics have raised the objection to the credibility of these accounts by pointing out that they may be due to religious indoctrination. However, toddlers or young children have not enjoyed the benefits of religious instruction and cannot possibly be aware of such matters.

Sam Parnia, a critical care and resuscitation specialist at NYU Langone Medical Center, published in 2014 the world's largest study of what happens to the human mind and consciousness in the early period around the time of death. Based on his research, Parnia believes that the conscious mind continues operating after the heart stops beating and the brain stops working for some period of time. Studies of near-death experiences are challenging the idea that our mind fades to black when our body expires.[2]

Telepathic communication is another example of how thoughts and feelings of one person can affect another. Imagine you are drinking a cup of coffee at Starbucks and suddenly, you have this feeling that someone is looking at you from behind. You turn around and you meet the eyes of the person who was looking at you. How did you apprehend that?

Or, consider the accounts of identical twins. Perhaps as many as one in five claims to share something mysterious: a special psychic connection. Identical twins are often very close, and share not only genes but home environments, friends, clothes and, of course, secrets.

I recall many years ago seeing a patient who had a twin brother living hundreds of miles away. One night, this man, let's call him Paul, suddenly experienced an overwhelming feeling of dread. Convinced it involved his twin brother, he rushed to the phone and called him. His wife answered and said that her husband, Paul's brother, was involved in a serious car accident. Hundreds of such occurrences have been reported. These are instances of telepathic communications that presently neither classical physics nor psychology can explain.

Lastly, there is the astonishing work of professor of medicine Ian Stevenson, published in six volumes by the University Press of Virginia, based on reports of hundreds of ordinary children who claim to have had past-life recollections. From this group he selected twenty cases in which the detail of recollections is so accurate and confirmed by multiple checks that in his opinion they represent genuine memories from previous lives. His paper on the past-life memories of children is *Children Who Remember Previous Lives.*

These are just a few examples of what is popularly referred to as *extrasensory perception (ESP)*[3] and what I call *sympathetic communication.* At this time the majority of classical neuroscientists reject these phenomena as pseudoscience, unproven New Age gibberish. I think they are wrong. I think sympathetic communication is real and well supported by research.

Where does sympathetic communication originate? No doubt in the brain and all its connected networks, the structures that support the embodied mind. Which brings on more questions. Is the mind separate from the body? What is it made of? Where does it dwell? Physical phenomena, at least at the macroscopic level, have a location in space and can be quantified. On the other hand, mental phenomena cannot be localized and cannot be quantified. The new generation of post-materialism scientists and those familiar with quantum physics have taken a run at this puzzling conundrum[4] that we shall presently explore.

Post-Materialism Science

Stuart Hameroff, from the Departments of Anaesthesiology and Psychology at the University of Arizona, emerged from obscurity in 1994 to advance what seemed—at the time—one of the more bizarre ideas about the human brain. Supported by Sir Roger Penrose, Hameroff suggested that quantum vibrational computations in *microtubules*, which are major components of the cell structural skeleton, were "orchestrated" ("Orch") by synaptic inputs and memories stored in microtubules, and baptised by Penrose "objective reduction" ("OR"), hence "Orch OR." They suggested that EEG rhythms (brain waves) derive from deeper level microtubule vibrations. **Most significantly, they further proposed that microtubules govern our consciousness.** This will not come as a total surprise to my attentive readers, who will recall that in the chapter on the cell, we already discussed the role microtubules may play in cellular memory.

Microtubules[5] are hollow, cylindrical structures, twenty-five-nanometer-wide tubes, thousands of times smaller than a red blood cell. They are found in every plant and animal cell (FIG 9.3). Microtubules provide internal support for living cells and act as conveyor belts, moving chemical components from one cell to another. During cell division, microtubules transport chromosomes from one end of the cell to the other, and then position the chromosomes in the new daughter cells.

Hameroff came to believe that microtubules play a defining role in consciousness. He points to the single-celled paramecium as evidence

(FIG 9.4). The paramecium, like the bacteria and slime molds we discussed, has no central nervous system, no brain, no neurons, but it swims around, finds food, seeks out a mate, and avoids danger. It seems to make choices, and it is definitely capable to process information. And since microtubules are *nanoscale* structures, Hameroff also began thinking that quantum physics might play a role in consciousness.

Hameroff suggests that consciousness arises from quantum vibrations in microtubules inside the brain's neurons, and connect ultimately to "deeper order" ripples in space-time geometry. "Consciousness is more like music than computation," he writes in *Interalia*, an online magazine dedicated to the interactions among the arts, sciences, and consciousness. Roger Penrose has remained committed to what the pair has co-published over the years.

New findings suggest that some of Hameroff's claims are more credible than previously assumed. Furthermore, the microtubule is suddenly a hot subject. And a growing number of scientists are saying that **quantum physics might be vital to our awareness, cognition, and even memory. In the last twenty years evidence has accumulated that proves the existence of quantum coherence in plant photosynthesis, bird brain navigation, and human sense of smell** (more on smell further on).

Physicist Neill Lambert of Japan's Advanced Science Institute researching photosynthesis has found supportive evidence that quantum effects can happen in biological systems at room temperature. That is to say, in our bodies.

Anthony Hudetz, a neuroscientist at the University of Michigan[6] who admits to have been quite skeptical of Hameroff's claims about microtubules in the past, based on recent evidence has declared himself ready to at least consider the possibility that Hameroff may be onto something important.

Riding this momentum, Hameroff and Penrose, together with a group of scientists from a variety of fields such as neuroscience, biology, medicine, psychiatry, and psychology, initiated a new science they call **Post-Materialist Science.** These scientists[7] emphasize that science is first and foremost a nondogmatic, open-minded method of acquiring knowledge about nature through the observation, experimental investigation, and theoretical explanation of phenomena. Its methodology is not synonymous with materialism and is not committed to any particular belief, dogma, or ideology.

I support this stand 100 percent. We should follow the evidence, rely on the data, but remember what Einstein once said, "Everything that can be counted does not necessarily count; everything that counts cannot necessarily be counted." This comment gains added significance and relevance as we are about to enter the realm of quantum physics also known as quantum mechanics, which we need to do to gain a deeper understanding of the brain-body-mind relationship.

Quantum Physics and the Rise of Quantum Biology

Having given much thought to the subject of this chapter, I have concluded that the only way to approach it is by gaining at least a rudimentary understanding of quantum physics also called quantum mechanics (QM). Therefore, I offer here a bare-bones introduction to quantum mechanics, with the caveat that I am not a physicist.

Quantum mechanics deals with the study of particles at the atomic and subatomic levels. Max Born coined the term in 1924. QM is complex, paradoxical, and hard to fathom if one is tethered too closely to classical Newtonian physics. So, dear reader, jettison your attachments to high school physics and take a walk on the wild side.

The theory sets fundamental limitations on how accurately we can measure particle locations and velocity, replacing classical certainty with probabilistic uncertainty. It describes just about every phenomenon in nature, both organic and inorganic, ranging from the color of the sky to the molecules and ions in living organisms. What makes quantum mechanics confusing is that the laws governing it differ drastically from classical physics.

A bit of historical background. During the 1920s and early 1930s, physicists discovered what has been called the wave-particle duality, a fundamental concept of quantum mechanics that proposes that elementary particles, such as photons and electrons, possess the properties of both particles and waves. What's even stranger is that the particle and wave aspects cannot be dissociated; rather, they complement one another. This is what Niels Bohr called the *principle of complementarity.* He saw this complementarity as the inevitable result of the interaction between a phenomenon and the apparatus used to measure it.

The mind compliments matter, just as the particle aspect of matter compliments its wave aspect. **Consciousness can interface with the material world because matter and energy are interchangeable.**

Quantum mechanics has confirmed that atoms and subatomic particles are not really solid objects—they do not exist with certainty at definite spatial locations and definite times. Matter is not solid in the way that it had been thought of since the rise of civilization.

The idea of atoms goes back to ancient Greece, where philosophers like Leucippus, Democritus, and Epicurus proposed that nature was composed of what they called ἄτομος (*átomos*), or "indivisible individuals." They thought that if you take an object—for example, a watermelon—and keep cutting it in half and then again in half into infinity, you will eventually end up with a particle that is so small that it is "uncuttable." Nature was, for them, the totality of discrete atoms in motion. Which, if you think about it, is downright brilliant considering they lacked any of the bells and whistles of modern science. Today, we know that the atom is almost entirely empty but for a swirling cloud of moving subatomic particles such as photons, electrons, neutrinos, quarks, etc. (FIG 9.5).

Remarkably, researchers discovered that particles such as electrons, being observed and the observer—the physicist, the apparatus and the method used for obscrvation—are linked. The scientists hypothesized that **the consciousness of the observer affects the physical events being observed, and that mental phenomena influence the material world**. Recent studies support this interpretation and suggest that **the physical**

world is no longer the primary or sole component of reality, nor that it can be fully understood without making reference to the mind.

Let's take a closer look. Early pioneers of quantum physics saw applications of quantum mechanics in the biological sciences. In 1944 Erwin Schrödinger discussed applications of quantum mechanics to biology. Schrödinger suggested that mutations are the result of "*quantum leaps.*" In 1963 Per-Olov Löwdin at Uppsala University proposed *proton tunneling* as another mechanism for DNA mutation. In his paper, he stated that there is now a new field of study called "*Quantum Biology.*"

Subatomic particles, atoms, or even entire molecules can exhibit *interference*, a classical property of waves in which two peaks reinforce each other when they overlap. Quantum effects such as interference rely on the *wave functions* of different entities being coordinated (they are said to be *coherent*) with one another. That sort of coherence is what permits the quantum property of *superposition*, in which particles are said to be in two or more states at once. If the wave functions of those states are coherent, then both states remain possible outcomes of a measurement.

Being in two states at once is not an unknown phenomenon in human psychology. Who has not had the experience of debating in their minds two contrary options such as, "Shall I write this letter of complaint or not?" One part of you says, "Give them hell!" and advocates in favor of writing the letter; the other cautions you, "Think of the consequences." This discussion can last a few seconds, minutes, or hours. Finally, you decide on a course of action. We often say, "I was of two minds" to describe this kind of a predicament.

You may have heard of lucid dreaming. In Eastern thought, cultivating the dreamer's ability to be aware that he or she is dreaming is central to both the Tibetan Buddhist practice of dream yoga and the ancient Indian Hindu practice of *yoga nidra*. For those unfamiliar with the term, a lucid dream is having a dream while asleep and developing the ability to control the dream in some way. The dreamer must let the dream continue but be conscious enough to remember that it is a dream. This can be achieved with preparation and practice. For example, you dream that a stranger is chasing you and you feel scared. Rather than give in to your anxiety and habitual pattern of fleeing, you turn around—still in your sleep—and confront the person. Doing so can be very therapeutic, especially if you have been a fearful person. Many psychotherapists use lucid dreaming as an integral part of therapy. The person who practices lucid dreaming is, at the time, really two persons or two minds. One is doing the dreaming while the other questions or directs the dreamer.

And, of course, there is the classic 1957 film about multiple personalities: *The Three Faces of Eve*. Suffering from headaches and inexplicable blackouts, timid housewife Eve White (Joanne Woodward) begins seeing a psychiatrist, Dr. Luther (Lee J. Cobb). He's stunned when she transforms before his eyes into the lascivious Eve Black, and diagnoses her as having multiple personalities. It's not long before a third personality, calling herself Jane, also appears. The film was based on a book by psychiatrists Corbett H. Thigpen and Hervey M. Cleckley, which in turn was based on their treatment of Chris Costner Sizemore, also known as Eve White.

Cases of multiple personality are rare today, but they are not unheard-of. They are listed in *DSM-5* of the American Psychiatric Association (2013) under Dissociative Identity Disorders. They are defined as "disruption of identity characterized by two or more distinct personality states." Once again, we are confronted by an enigma and need to turn to quantum physics for a plausible explanation.

If one quantum particle interacts with another, they connect and become linked into a composite *superposition*: in a sense, they become a single system. The two objects are then said to be *entangled* (FIG 9.6). Einstein, no friend of quantum mechanics, referred to entanglement as "spooky action at a distance."

Entangled particles are intimately joined since the day they were created. No matter what distance separates them, be it the width of a lab table or the breadth of the universe, they mirror each other. Astonishingly, whatever happens to one instantaneously affects the other and vice versa.

Jian-Wei Pan, physicist at China's University of Science and Technology, dramatically demonstrated this recently in a new study. Pan and his team produced entangled photons on a satellite orbiting 300 miles above earth and beamed these particles to two different ground-based labs 750 miles apart, all without losing the particles' strange linkage. The previous distance for what's known as *quantum teleportation* or sending information via entangled particles, was about 140 kilometers, or 86 miles.

At this time scientists still can't explain how the particles are separate but connected. Does that not remind you of twins communicating with each other "telepathically" or, to use my term, "sympathetically" or, of the other cited examples of sympathetic communication?

Let's return to biology and something familiar: the sense of smell. Up until the late twentieth century the accepted theory of olfaction was based on the shape of the molecules of the substance emitting the odor and how they docked with their specific receptors. This was commonly referred to as the "lock and key" hypothesis. Then in 2006, Luca Turin, presently visiting professor in theoretical physics at the University of Ulm, published his vibration theory in *The Secret of Scent: Adventures in Perfume and the Science of Smell*, and all hell broke out within the scientific community.

Turin proposed that scent is transmitted by vibration, and that the human nose is so engineered that it is able to process these vibrations and interprets them as smells, ultimately leading to olfactory perception. Through an operation called *electron tunneling* receptors in the lining of the nose unravel their coded content. Furthermore, each molecule's pattern of vibration—think of musical notes—plays the same way in every person's nose.

There are 390 functional olfactory receptors in humans that can respond to 100,000 or more odorants, thus eliminating the concept of 1:1 receptor to odorant matching. Olfactory receptors are versatile and able to respond to chemicals never encountered before. I call that good planning by evolution or a higher power.

The old model of lock and key is now being replaced by what Jennifer C. Brookes of University College London has termed the "swipe card" model of odorant recognition. It proposes that while the shape must be good enough, other information characterizing the odorant is also important.

In this case the additional information on the "card" is the **vibration frequency of the molecule** in the odorant.

According to Brookes, Turin's proposal of vibration frequencies monitored by inelastic electron tunneling stands up well to scrutiny since vibration frequency is a crucial part that dominates smell. In her opinion the swipe card description is a more useful paradigm than lock and key.

Simply put, the **mechanisms underlying olfaction involve quantum processes**. Although still in the experimental stage and not yet proven, it seems highly likely that the same or similar processes operate in other receptors activated by small molecules such as neurotransmitters, hormones, steroids, and so on. **This provides added credence to viewing our bodies and minds as being governed by a confluence of classic and quantum laws of physics.**

Stuart Alan Kauffman, theoretical biologist and complex systems researcher currently emeritus professor of biochemistry at the University of Pennsylvania, with Samuli Niiranen and Gabor Vattay, was issued a founding patent[8] on the *Poised Realm*, an apparently new "state of matter" hovering reversibly between quantum and classical realms, between quantum coherence and classicality.

Kauffman thinks that the system seen in the chlorophyll molecule (which he studied at length) raises the possibility that webs of quantum coherence or partial coherence can extend across a large part of a neuron, and can remain poised between coherence and decoherence. **Kauffman believes that this Poised Realm in the human brain is where consciousness reigns.** And consciousness is surely one aspect of the mind.

How does all of this relate to our understanding of the nature of the mind?

Theories of the mind fall into four general categories.

1. The mind is separate from the brain and is not controlled by the laws of physics of any kind. This is essentially a religious or spiritual view that assumes that the mind has

been in the universe all along. Individual minds are parts of a greater all-encompassing mind that may or may not be God.

2. The mind is the product of complex neuronal activities of an individual's brain. It is an evolutionary adaptation. In scientific jargon, "the mind is an epiphenomenon of the brain," an emergent quality; in other words, a by-product of a functioning brain.
3. The mind obeys physical laws not yet fully understood, acting on the neurons of the cerebral cortex. This is the view proposed by the philosopher Albert North Whitehead and elaborated on by Hameroff. In his theory, as I understand it, consciousness and the mind are epiphenomena of quantum computations in brain microtubules. To Hameroff the mind is an intrinsic factor of the universe. To me this seems to represent a marriage of theories #1 and #2 above.
4. Finally, there is the Kauffman hypothesis, according to which the mind, consciousness, and free will are associated with the Poised Realm. Our brains with our sense organs connect us to the universe. The difference in theories between Hameroff and Kauffman is that the former locates consciousness in the microtubules and the latter in the Poised Realm. They both lean heavily on quantum physics for their hypotheses.

We know that the mind can influence the state of the physical world. We know that the intentions, emotions, and desires of an experimenter may influence experimental outcomes, even in controlled and blinded experimental designs. We have explored subjects like general anesthesia, hypnosis, placebos, near-death experiences (NDEs), and out-of-body experiences (OBEs), which separately and jointly establish that the **brain acts as a transceiver of mental activity; i.e., the mind can work through the brain, but is not necessarily produced by the brain.**

When we sleep, we are not fully conscious but we are also not fully unconscious, as for example lucid dreams prove. If a person is under general anaesthesia, should one of the doctors say something alarming like, "Oh, I think we just ruptured her stomach," the patient will react with all the physical signs of a panic attack. The brain maybe underperforming but the mind is fully operational.

Summary

Materialism is an attractive philosophy; at least it was before quantum mechanics altered our thinking about matter. Quantum mechanics has revolutionized the study of physics and biology and opened new horizons. The history of science has taken us from our physical senses and "common sense" to "quantum non-locality" and "spooky action at a distance"—and in the process revealed an ever more baffling reality. Furthermore, quantum mechanics has shown that human thoughts, intentions, and emotions may directly affect our material world.

Today mathematicians and physicists generally accept the fact that the world is fundamentally governed by quantum rules. There is no need, they say, to regard Newtonian physics and quantum physics as mutually exclusive, since they are not only consistent but also inextricably linked.

Post-materialism science is pointing the way toward expanding our psychic space. Hamerroff's and Kauffman's ideas have generated much excitement among researchers toiling at the interfaces between physics, mathematics, biology, information theory, psychology, and philosophy, where mysteries and paradoxes abound. Equally significant are the contributions of Turin, Brookes, Charles Sell, and others on olfaction, showing that it is partly a result of quantum operations.

In view of recent research on quantum biology, microtubules, and the Poised Realm, particularly by the phenomena of entanglement and non-locality, both telepathy/sympathetic communication and telekinesis are possible. More important, quantum biology goes a long way in explaining how the mind affects matter, which we know it does—think placebo effect, psychosomatic medicine, and the like.

The view that the mind is both dependent and independent of the brain and the rest of the body is supported by experimental evidence from the very cutting edges of academic scholarship. Like protons or electrons that, depending on circumstance, can be particles or waves and everything in between.

All my professional life, be it by temperament, upbringing, education, and a host of other influences, I have embraced the deterministic, hard scientific view of the world. However, moved by the research on which this chapter is based, I find myself ready to accept the for me radical possibility that perhaps a precursor of the mind not yet understood may be one of the fundamental elements such as mass, gravity, or electric charge that the world is made of.

The hypothesis I subscribe to is that consciousness and free will are qualities of the mind and that the mind is more than the brain.

KEY TAKEAWAYS

- **General anesthesia does not shut down the brain globally and does not always produce a complete absence of consciousness.**

- **Our current understanding of matter alone is unlikely to explain the nature of mind.**

- **Orch OR may provide a bridge between neuroscience and more spiritual approaches to consciousness.[9]**

- **Post-materialist science suggests that there is a connection between the brain's biomolecular processes and the basic structure of the universe.**

- Quantum coherence has been demonstrated to be present in plant photosynthesis, bird brain navigation, and human sense of smell.

- The mechanisms underlying olfaction involve quantum processes. Although still in the experimental stage, it seems reasonable to assume that the same or similar processes operate in other receptors activated by small molecules such as neurotransmitters, hormones, steroids, and so on.

- Our bodies and minds are governed by a confluence of classic and quantum laws of physics.

CHAPTER TEN

SIGNIFICANCE

I believe that only through the relentless scrutiny, questioning, and challenging of orthodoxies can science progress. The information that appears on the preceding pages is an internally coherent framework that synthesizes empirically supported research from diverse fields of investigation. It challenges many of the prevalent tenets of neuroscience. But it does more than that. The Embodied Mind Hypothesis will benefit you on a personal as well as on a societal level.

This paradigm-shifting embodied mind hypothesis will improve your life in three distinct ways. Firstly, as the scientific community, especially the medical fraternity and pharmaceutical industry, gradually translate these ideas into practice they will discover new and better treatments of many physical and mental diseases.

Secondly, a large number of the studies reviewed in this book deal with the mysteries of memory. We know that it is not only the brain that is the keeper of our past experiences. Rather, memory is embedded in the matrix of our bodies. It is everywhere. Which means that one day in the case of brain tumors or Alzheimer's or similar neurological diseases there will be therapies capable of retrieving memories from the body and returning the person to normal—or at least vastly improved—functioning, very much like the French civil servant I mentioned before whose body performed this task on its own.

The attentive reader, now familiar with epigenetics and how the lives of parents and grandparents several generations back can affect their children even from before conception, will keep this information in mind and adjust their lifestyle accordingly should they plan to have children. Readers past that stage of life may be able to come to resolve a variety of emotional issues by tracing their origins to their in-utero life and the life experiences of their ancestors. Such insights can be truly healing.

Lastly, knowing fully well the incredible power that thoughts exert over our bodies the reader will take a deep dive into their own minds, identify any self-defeating, self-sabotaging scripts hiding in the dark corners of their unconscious and ditch them forever. Of course, this will take time and effort; perhaps even the support of a therapist, psychologist, or psychiatrist. But think of this as no different than seeing another medical doctor for a physically manifesting ailment. Remember, perseverance will pay off.

Health and Disease

Currently, care of sick patients is often restricted to separate disciplines. Neurologists focus on disorders of the brain and nervous tissues. Cardiologists deal with diseases of the heart. The same attitude has been also widely adopted in research. However, such an organ-centered, blinkered perspective does not take into account the multiple body systems described on the preceding pages that operate simultaneously and in synchrony. This is where changes in practice will need to be made.

A case in point is the well-known fact that patients suffering from heart failure have an increased risk to develop age-associated dementia. Now, researchers from the German Center for Neurodegenerative Diseases have found a possible cause for the increased risk of dementia in people with heart problems. In their experiments, mice afflicted with heart failure developed cognitive decline. This in turn was linked to increased cellular stress pathways and altered gene activity in neurons of the hippocampus.

These pathological changes were ameliorated by the administration of a drug that promotes neuronal cell health.

Based on these discoveries, it is not surprising that experts from the German Center for Neurodegenerative Diseases call for interdisciplinary approaches and novel therapeutic avenues to the treatment of neurocognitive disorders.

Among neurocognitive diseases the most common and most dreaded by older people is Alzheimer's. Recent estimates indicate that Alzheimer's disease may rank third, just behind heart disease and cancer, as a cause of death for older people in the United States. Alzheimer's patients' brains contain many abnormal clumps (amyloid plaques) and tangled bundles of fibers (neurofibrillary, or tau, tangles). Most therapies designed to treat Alzheimer's target theses plaques, but without much success.

Pharmaceutical companies understandably have focused their efforts on brain pathology. In doing so, they have overlooked the fact that the brain is connected to the body by multiple two-way communication systems. Through these channels a change in any part of the body will affect the brain and vice versa. The brain, as I have said before, does not work on its own.

One of the brain's main inputs is from the microbes in our gut. The microbiome has multiple, critical effects on our physiological and metabolic processes ranging from prenatal brain development and modulation of the immune system to, perhaps most surprisingly, behavior and cognition. This microbiome communicates with the enteric nervous system ("The Thoughtful Bowel"), the immune system, and by way of the vagus nerve with the brain. It needs to be stressed that the vagal gut-to-brain axis plays a critical role in motivation and reward. Along these routes exchange of information takes place in both directions. These linkages represent further essential constituent parts of the Embodied Mind.

In a new study, researchers at the University of California San Diego School of Medicine, linked the gut microbiome to social behavior, including findings that people with more diverse microbiota tend to have larger social networks and tend to be wiser. They observed that reduced microbial diversity typically is associated with a variety of diseases, including obesity, inflammatory bowel disease, and major depressive disorder.

The relationship between loneliness and microbial diversity was particularly strong in older adults suggesting that older adults may be especially vulnerable to health-related consequences of loneliness. At this time, we do not know which comes first: Does loneliness lead to changes in the gut microbiome or, reciprocally, do alterations of the gut environment predispose an individual to become lonely?

Scientists at St. Jude's Children's Research Hospital are studying how signals that muscles send when stressed affect the brain. The researchers found that stress signals rely on an enzyme called *Amyrel amylase,* and its product, *maltose.* The scientists showed that the release of this enzyme promotes protein quality control in distant tissues like the brain and retina. The signals work by preventing the buildup of misfolded protein molecules in neurons. Findings suggest that synthetic reproduction of the signaling enzyme may potentially help combat neurodegenerative conditions like age-elated dementia and Alzheimer's disease.

But it is not only muscles that send messages to the brain. Now we are discovering that so do bones. You will recall that in the chapter on the wisdom of the body we learned that there is more to the stress response than cortisone and adrenaline. It turns out that *osteocalcin*, a hormone made in bones, regulates a large and continually growing number of physiological functions and developmental processes. Numerous studies have found that it helps regulate metabolism, fertility, brain development, and muscle function. It improves memory and lowers anxiety and is linked to the biology of aging.

James Herman of the University of Cincinnati has written that chemical messages from other parts of the body may also play a role in the stress response. Experiments from his own lab have identified a potential stress-related role in the stress response. Experiments from his own lab have identified a potential stress-related role for signals secreted from adipose tissue, the loose connective tissue, located beneath the skin.

All of the above studies illustrate this work's fundamental principle: no organ whether it be the brain, kidney, heart, or any other is an island unto itself.

Mysteries of Memory

Studies in epigenetics are particularly relevant to this work. In high school we were taught that genes make us who we are. Wrong. Gene expression does. And gene expression varies, depending on the life we live. Equally important, before their offspring are even conceived, parental life experiences and environmental exposures modify their germ cells and in turn affect the development and health not only of their children, but even of their grandchildren and great-grandchildren. Epigenetic mechanisms allow organisms to adapt to environmental changes in a very short time, sometimes instantly, rather than over millennia as is the case with Darwinian evolution.

There is now robust biological evidence of transgenerational transmission of trauma to offspring by both parents. Mothers pass on trauma and other negative emotions like anxiety and depression by way of their insulin-like proteins and modification of their unborn child's hypothalamic-pituitary-adrenal axis (HPA axis) and fathers through their extracellular vesicles. In addition, probable familial factors previously discussed are micro RNAs (miRNAs) and lncRNAs in the circulation as well as epigenetic changes in maternal and paternal germ cells.

What makes these findings so relevant is that they demonstrate that our bodies, and by this I mean our whole body, including the brain, is constantly in a state of flux, constantly changing. Whether that change is initiated internally or externally, when one element changes, all other parts of the body including the brain are affected.

We have seen that ova and sperm pass on at conception not only instructions for building a child's body but also for constructing their mind. We encounter here for the first time the incredible wealth of information carried in cells.

In spite of their minuscule size, these tiny biological machines are surprisingly intelligent. Cells communicate with neighboring cells and provide them with aid in case they suffer from stress or sickness. Cells remember their origins, all the way back to conception.

The cell membrane, which possesses all the characteristics of microchips, in addition to performing other duties, also directs the production of proteins that encrypt memories. This would apply to all cells in the body, in other words, to both cortical cells and somatic cells.

DNA in the cell nucleus has been shown to be a compact and stable information-storage medium in biology and computer technology. Once again, it needs emphasizing that DNA has the potential to store a massive amount of data. At the present, we know what some of this information is but much of it remains a mystery. It is like the dark side of the Moon. Even if we can't see it we know it exists. It is in this unexplored area of DNA, in addition to the cellular membrane and the cytoskeleton that I hypothesize memories of trauma, fears, phobias, or special attributes or talents are cached.

Over the past four hundred years, the discovery of the telescope ushered in a new appreciation of the vastness of the universe as each successive generation of astrophysicists realized that the universe is bigger than the previous generation thought. With the introduction into biology of the electron microscope a similar deepening of understanding of the human cell occurred. Here, we have discovered relatively huge spaces within which the nucleus, with its DNA and RNA, fifteen tiny organelles, and forty-two million protein molecules live and work. And now, we are expanding our knowledge of atoms and finding once again more open spaces, more complexity, and more hidden surprises.

All of the major mechanisms by which nerves function—neurotransmitters, ion channels, and electrical synapses are utilized by all the cells in the body. There is no mistaking the fact that processes and activities thought to represent "recent" evolutionary specializations of the nervous system represent ancient and fundamental cell survival processes that still persist in all our cells today.

Thus, biological cells in tissues and organs work in concert with neural networks and represent the third system comprising the Embodied Mind. However, we must not forget about the biological cells in the circulatory system. These are made up of the red blood cells and white blood cells. In the latter group are the cells of the immune system.

Scientists everywhere agree that the memory cells of the immune system collect a treasure trove of information on the predators they encounter over a lifetime so they can fight them should they reappear. Does it not make sense to at least consider the possibility that cells as clever as these immune cells might not also contain fragments of life memories, especially since reliable evidence points to the existence of a complex feedback loop between the immune system and the brain? Thus, the immune system jointly with the other systems already described comprises the embodied mind.

At the beginning of this chapter, I pointed out the importance to our health of the microbes in our gut. Biologists have known for decades that bacteria can use chemical cues to coordinate their behavior. The best example, elucidated by Bonnie Bassler of Princeton University and others at the beginning of this chapter, is quorum sensing, a process by which bacteria extrude signaling molecules until a high enough concentration triggers cells to form a biofilm or initiate some other collective behavior.

But Gürol Süel, biophysicist at the University of California San Diego, and other scientists, are now finding that bacteria in biofilms can also talk to one another electrically. Like neurons, bacteria apparently use potassium ions to propagate electrical signals. We are learning about an entirely new mode of communication among bacteria. Bacterial genesis geneticist James Shapiro supports these findings. He has argued that bacterial colonies might be capable of a form of cognition.

If colonies of bacteria may be capable of a form of cognition would the cells in our bodies, which are much more complex than bacteria, not do as well or better? In fact, there is mounting evidence from many sources that show that all body cells and tissues, not just neural cells, store memory, inform genetic coding, and adapt to environmental changes, thus forming the bedrock of the embodied mind. Cellular memory is not just an abstraction but the well-documented scientific fact that shifts our understanding of memory.

Close cousins to the cells in our bodies are unicellular organisms such as bacteria and slime molds. While containing all the structures of the "ordinary" cells and like them have no neurons or brains, they are capable of reproducing and learning, decision-making, goal-directed behavior, and memory. They sense and explore their surroundings, communicate with their neighbors, and adaptively reshape themselves.

How does the brain fit into the embodied mind? Received wisdom held until recently that synapses keep memories. This is no longer true. Recent research has overtaken the synapse theory, which has been inching toward a slow death for years, as generational changes happen in the scientific academy.

Memory engrams are stored in neurons. And of course, each neuron has one axon and several thousand dendrites. The growth cones of axons contain much of the proteins involved in growth, metabolism, and signaling while the tiny compartments in dendrites can each perform complicated operations in mathematical logic. Truly astonishing capabilities.

Neurons in the cerebral cortex are supported by glial cells, the most significant of which are the astrocytes. Astrocytes span vast numbers of neurons and millions of synapses and seem to contribute another level of functioning to neural networks. Some researchers have spoken of astroglia nets as a potential information processing system. This is significant because it demonstrates the fact that neurons are not the only cells in the body that do the heavy lifting when it comes to thinking and memory.

If the seat of all memory were truly the brain, as old-school scientists still claim, then to ensure long-term storage of information, the brain cells and their circuits would need to remain stable and intact over a lifetime. Yet this is not the case in the animals we discussed in the chapter on regeneration, hibernation, and metamorphosis, where we saw that memories survive drastic cellular turnover and rearrangement.

A good illustration of this are planaria, which contrary to the prevalent scientific view, can develop into "new" individuals from small portions of their "old" bodies and remember what they learned before they were truncated and lost their heads. Of particular relevance here is Shomrat and Levin's research on planaria that led them to conclude that traces of memory of the learned behavior are retained outside the brain. They suggested that this was by way of mechanisms that include the cytoskeleton, metabolic signaling circuits, and the gene regulatory networks. This loss and rearrangement of brain neurons is observed in hibernating animals and is even more pronounced in metamorphosing animals.

What accounts for how larval experience can persist through pupation into adulthood—in other words, the survival of these animals' memories from their early embryonic stage (eggs, larvae)—long before they had a mature brain? The only credible answer to that question, based on the research presented, seems to be that memory, in addition to being stored in the brain, must also be encoded in other cells and tissues of the body. The brain never works alone. We are all endowed with both somatic and cognitive memory systems that mutually support each other.

People who have had parts of their brain tissues surgically removed or were born with water on their brains—hydrocephaly—in many instances have been noted to perform quite normally, often unaware of their handicap. I believe this is not because of "neuroplasticity" or "recruitment" of unaffected areas in the brain but because the brain, or whatever is left of it, is—and always was—part of a much larger communication network in the body. This extended brain, the embodied brain acts as a backup memory storage system and "tops up" whatever data are missing in the brain.

Humans are not as versatile as planarian worms or the animals that hibernate or undergo metamorphosis. However, our bodies are perfectly capable to heal wounds or broken bones or even replace a complex structure such the liver. Repair of damaged tissues or organs is not top-down but bottom-up, not controlled by the brain but organized locally by the affected cells or assemblies of cells. It is this ability of "remembering" the body's own structure by every cell in the body that makes healing and regeneration of body tissues possible.

Traumas, such as PTSDs, whether physical or psychological, or highly emotionally charged experiences, are locked in the whole body. Occasionally, our bodies speak loudly about things we would rather not hear. That is the time to pause and listen.

The one organ in our bodies that we really should pay more attention to is our heart. There exist more expressions and metaphors about the heart than of any other organ in our body. In common parlance the heart is often spoken of as a center of thought, feeling, and personality. Yet we hardly ever consider these as reflecting scientifically valid observations. It turns out these expressions are closer to recent discoveries in cardiac function than anyone in academia previously assumed.

Like the digestive tract but even more so, the heart is connected to the brain by the vagus nerve and the autonomic nervous system. The heart is also part of the body's endocrine system and secretes hormones of its own. Furthermore, it generates a unique electromagnetic field that affects the rest of the body and extends several feet beyond.

The signals the heart sends continuously to the brain influence the function of higher brain centers involved in perception, cognition, and emotional processing, and vice versa. Because of this feedback loop, it is not surprising that people who suffer from depression or schizophrenia, or simply too much stress, have been shown to be at a higher risk to develop heart disease than emotionally healthy individuals.

The heart's brain consists of an intricate network of several types of neurons, neurotransmitters, proteins, and immune cells similar to those found in the enskulled brain. This *heart brain* enables the heart to act independently of the cranial brain—to learn, remember, and even feel and sense. Considering the evidence for information/memory being stored in all the cells in our bodies—the organizing principle of this book—it seems reasonable to assume that cardiac cells would be no exception to that, and if anything, more likely to carry data of a personal nature. Consequently, it is not surprising that sensitive transplant patients may manifest personality changes that parallel the experiences, likes, dislikes, and temperament of their donors.

We know that life is built on the interactions of signaling entities: communities, people, organs, cells, bacteria, and even viruses. As documented in this book, the capabilities of our body cells have been underestimated for far too long. Whether we are conscious of it or not, all the cells in our bodies—neurons, immune cells, and somatic cells in concert with the tissues of the heart, our genome, the enteric nervous system, and the microbiome form networks that connect to the enskulled brain and form the seedbed for the Embodied Mind. All memories, consciousness, and the mind emerge from this dynamic, linked sentient network.

Mind Power

In the chapter on the immune system, I briefly referred to studies that show the effects of meditation on telomeres. As you may recall, telomeres protect the ends of chromosomes by forming a cap, much like the plastic tip on a shoelace. Without telomeres our chromosomes could end up sticking to neighboring chromosomes and could not function. The telomeres are longest at birth and shorten with increasing age. Once a cell's DNA loses its telomere, it can no longer divide and dies. Massive cell deaths bring about aging. Consequently, telomere length is commonly considered as a potential biomarker of aging and age-related diseases. Telomeres like immune cells respond to emotions. Negative emotions shorten telomeres, while happy feelings help to maintain their integrity.

A study at Charité—University Medicine Berlin on telomeres takes this subject a step further. In a sample of 656 mother-child dyads the investigators conducted serial assessments over the course of pregnancy to quantify maternal stress, negative and positive emotional responses to pregnancy events, positive affect (mood), and perceived social support. What they found was that *maternal stress* significantly predicted shorter newborn telomere length and *maternal resilience* was significantly associated with longer newborn telomere length. For the purposes of this study, *resilience* was defined as a multidimensional measure that incorporates positive affect and perceived social support.

Resilience may also exert a protective effect on telomere length by way of the immune system. Positive psychological states and social support have been associated with lower levels of cytokines and inflammatory markers and a reduced risk of infection.

Resilience, positivity, and social support are known to diminish stress-related autonomic arousal and lead to a more rapid and complete recovery from stress. This beneficial effect of resilience highlights the importance of attending to mothers' physical and mental health during pregnancy to optimize both hers as well as her child's health. Naturally, this applies not only to pregnant mothers but also to the rest of us. There is much research to support the dictum that good health is fostered by social support, by face-to-face interactions with extended family and friends, participating in social clubs and/or houses of worship, engaging in volunteer work, and enjoying the companionship of fellow workers. These factors contribute to one's feelings of belonging, of being part of humanity, and those feelings are the best antidotes to stress.

At the time I was writing this, a study from Oregon State University on aging came to my attention. It found that how people thought about themselves at age fifty predicted a wide range of future health outcomes up to forty years later—cardiovascular events, memory, balance, will to live, hospitalizations, even mortality. This research reminds me of the work of Steve Cole at the UCLA School of Medicine, who I cited in the first chapter on genetics. You may recall that Cole published much on the subject of self-regulation. He holds, and I totally agree with him, that we can be architects of our own lives if we only make a conscious effort to do so.

Turning our attention to the age-old brain-mind conundrum now, sleep, lucid dreaming, hypnosis, placebos, sympathetic communication, and general anesthesia have all been scrutinized to explore consciousness and the mind. We know that *anesthetic agents* do not blunt our brain function globally but exert specific and dose-dependent effects on brain systems that sustain internal consciousness and perception of the environment. Each agent induces distinct altered states of consciousness. Anesthesia is clearly different from physiological sleep but also shares many similarities with it. Particularly, in the sense that emotionally charged words, like, "Oops, I think I nicked his ulnar nerve," will have a physiological effect on the person, though they may not remember it afterward. Which means that the mind is always active. It does not take holidays. It is always on call. The brain may be asleep but the mind is awake.

Hypnosis has been successfully used to induce anesthesia without drugs. Hypnosis, also referred to as hypnotherapy or hypnotic suggestion, is a trancelike state in which one has heightened focus and concentration and is more open to suggestions. Usually, a therapist using verbal repetition and mental images induces hypnosis.

Hypnosis may alleviate pain due to burns, cancer, childbirth, irritable bowel syndrome, fibromyalgia, temporomandibular joint problems, dental procedures, and headache. Hypnosis can be used to help you gain control over undesired behaviors or to help you cope better with anxiety or pain. Hypnosis has been beneficial in the treatment of insomnia, bed-wetting, smoking, and overeating. Hypnosis has been valuable in easing the side effects related to chemotherapy or radiation. It's important to know that although you're more open to suggestion during hypnosis, you don't lose control over your behavior. Your mind is still looking out for you.

In a normal waking state, different brain regions share information with each other. Researchers from Finland's University of Turku found that during hypnosis the brain shifts to a state where individual brain regions act more independently of each other, and the various brain regions are no longer synchronized. Essentially, hypnosis alters the way the individual processes information, which allows the mind to effect physiological changes in the body as well as personality changes. Your mind can be a powerful healing tool when given the chance.

A prime example of that, following in the footsteps of hypnosis, is the well-known *placebo effect*. The term *placebo* refers to an inert treatment or substance that has no known effects, while the term *placebo effect* refers to the effects of taking a placebo that cannot be attributed to the treatment itself. When this response occurs, many people have no idea they are responding to what is essentially a "sugar pill."

The fact that the placebo effect is tied to a person's expectations doesn't make it imaginary or fake. Placebos have been shown to produce measurable physiological changes, such as an increase in heart rate or blood pressure. However, illnesses that rely on the self-reporting of symptoms for measurement are most strongly influenced by placebos, such as depression, anxiety, irritable bowel syndrome, and chronic pain.

Verbal, behavioral, and social cues contribute to a person's expectations of positive or negative results. The placebo effect involves complex neurobiological reactions that include increases in feel-good neurotransmitters, like endorphins and dopamine, and greater activity in certain brain regions linked to moods, emotional reactions, and self-awareness.

All of it can have therapeutic benefit. "The placebo effect is a way for your brain to tell the body what it needs to feel better," says Ted Kaptchuk of Beth Israel Deaconess Medical Center, whose research focuses on the placebo effect.

The attention and emotional support you give yourself is often not something you can easily measure, but it can help you feel more comfortable in the world, and that can go a long way when it comes to healing.

It follows that it is time to move the dial on the scientific clock and seriously dispute the commonly accepted deterministic view of the mind as an epiphenomenon of the brain. Instead, I propose we entertain the concept of the embodied quantum mind, in which the mind is both dependent on and independent of the brain and the rest of the body. Like protons or electrons that, depending on circumstances, can be particles or waves or the in-between states of wavicles, the mind is fluid and adaptive.

Communication between neurons or any other cells in the body involves the movement of electrically charged atoms called ions from one nerve cell to another in order to generate outputs such as the movement of a finger or thinking of a sentence for this page. While these cells contain and pass information, I don't think we can attribute to them consciousness or a mind. It is only when they unite their resources that the mind and consciousness arise just like we see happening, on a more primitive level, with bacterial films.

Remember Albert Einstein's equation, $E = mc2$. It states mathematically a law of physics according to which matter can be changed into energy.

Recently, scientists at Imperial College London think they've figured out how to turn energy directly into matter. Both energy and matter are physical, but while matter encodes information as discrete particles separated in space, energy information is encoded as overlapping fields in which information is bound up into single unified wholes. It is a form of dualism, but a scientific dualism based on the difference between matter and energy.

The Embodied Mind is both material and nonmaterial, depending on the presence of an observer. When we are in a state of self-reflection or relating to another person, we act as observers. Introspection opens the portals to access the wisdom of the Embodied Mind.

The studies referenced on these pages and many more in the literature show that consciousness and the mind, functions traditionally attributed to the brain alone, exist as part of the body-wide-web of systems, including but certainly not limited to the brain. The mind is fluid and adaptable embodied but not enskulled.

The Embodied Mind hypothesis allows for the agency of free will and empowers each of us to fully live up to our potential by way of self-regulation rather than by the exigencies of an unpredictable environment.

ACKNOWLEDGMENTS

Writing this book has been a long and at times difficult journey. I could not have completed it without the constant encouragement and loving support of my wife Sandra, as well as of my children David and Lynette, and grandchildren Alec, Sidney, James, and Elkah. Lorraine Tadman, my hardworking research assistant, did an incredible job editing, formatting, and wrestling to the ground endnotes and references, for which I am deeply grateful.

I want to thank friends Michael McKenna, Rick Graff, Brendan Howley, Peter Samu, and Sy Silverberg, who took the time to review various chapters and offer suggestions. I am grateful to my outstanding literary agent, Don Fehr, who provided essential critiques and sage guidance during the preliminary stages of writing the book proposal. Once satisfied, he successfully placed the manuscript with Pegasus in New York.

This turned out to be an excellent fit. I loved working with the publisher, Claiborne Hancock, and especially with my editor and deputy publisher, Jessica Case. I very much appreciate her assistance, patience, and skill.

Needless to say, the opinions expressed in this book are mine alone, not theirs. I am the responsible party, not they.

ENDNOTES

Chapter 1. Do Genes Matter?

1. *morphogenesis* (page 9): The shaping of an organism by embryological processes of differentiation of cells, tissues, and organs and the development of organ systems according to the genetic "blueprint" of the potential organism and environmental conditions. https://www.britannica.com/science/morphogenesis.
2. *Concerning 'Bias'* (page 33): Two African hives ("killer bees") and two European hives (a gentle bee strain). Please, I am not trying to offend anyone here. The study is about bees, not people.
3. *cognitive flexibility* (page 16): Cognitive flexibility refers to the ability to shift attention between task sets, attributes of a stimulus, responses, perspectives, or strategies. Perone, S., Almy, B., and Zelazo, P. D. "Toward an understanding of the neural basis of executive function development." In *The Neurobiology of Brain and Behavioral Development* (291–314). (Cambridge, Mass.: Academic Press, 2018).
4. *HPA axis* (page 26): The hypothalamic pituitary adrenal (HPA) axis is our central stress response system. The HPA axis links the central nervous system and endocrine system in a self -reinforcing loop.

Chapter 2. The Brain: How It Remembers What It Remembers

1. *cerebellum* (page 40): In addition to the cerebral hemispheres our skulls also contain a small structure called the *cerebellum* that sits at the back of the brain underneath the two hemispheres. It looks different from the rest of the brain because it consists of much smaller and more compact folds of tissue. It represents about 10 percent of the brain's total volume but contains 50 percent of its neurons. In the past, the cerebellum was seen to control mainly postural and motor balance. It now appears that the cerebellum is a critical neuromodulator also of intellect and mood, optimizing these functions that are represented with topographic precision in distinct regions of the cerebellum. Most important, the cerebellum serves as a quality check on movement; it also checks our thoughts as well—smoothing them out, correcting them, perfecting things. In the past few years evidence has accumulated that indicates prominent cerebellar dysfunction within the cerebello-thalamo-cortical networks in schizophrenia. Stoodley, C. and Schmahmann, J. (2009). "Functional Topography in the Human Cerebellum: A Meta-analysis of Neuroimaging Studies." *NeuroImage,* 44(2), 489–501. doi:10.1016/j.neuroimage. 2008.08.039; Demirtas-Tatlidede, Asli, Freitas, Catarina, Cromer, Jennifer R., et al. (2010). "Safety and Proof of Principle Study of Cerebellar Vermal Theta Burst Stimulation in Refractory Schizophrenia." *Schizophrenia Research.* 124(1–3): 91–100.

2. *IQ* (page 41): The vast majority of people in the United States have IQs between 80 and 120, with an IQ of 100 considered average. To be diagnosed as having mental retardation, a person must have an IQ below 70–75, i.e., significantly below average.
3. *Mediterranean diet* (page 44): The Mediterranean diet is not a "diet" per se. It is a mix of the traditional eating habits of people living in Spain, Italy, France, Greece, and the Middle East.
   How to Start the Mediterranean diet?
   Eat natural, unprocessed foods like fruits, vegetables, whole grains, and nuts.
   Make olive oil your primary source of dietary fat.
   Reduce the consumption of red meat (monthly).
   Eat low to moderate amounts of fish (weekly).
   Drink a moderate amount of wine (up to one to two glasses per day for men and up to one glass per day for women).
4. *(MRI) scan* (page 45): MRI scan is a medical test that uses a magnetic field and radio waves to create a detailed picture of organs and other structures inside the body. MRI stands for magnetic resonance imaging.

3. The Immune System: This Doctor Makes House Calls

1. *by producing antibodies* (page 49): Antibodies are produced by the immune system in response to the presence of an antigen. Antigens are large molecules, usually proteins, on the surface of cells, viruses, fungi, bacteria, and some nonliving substances such as toxins, chemicals, and foreign particles. Any substance capable of triggering an immune response is called an antigen.
2. known as *T cells* (page 49): figure 4. T Cell Distribution. *Credit: L. Tadman*

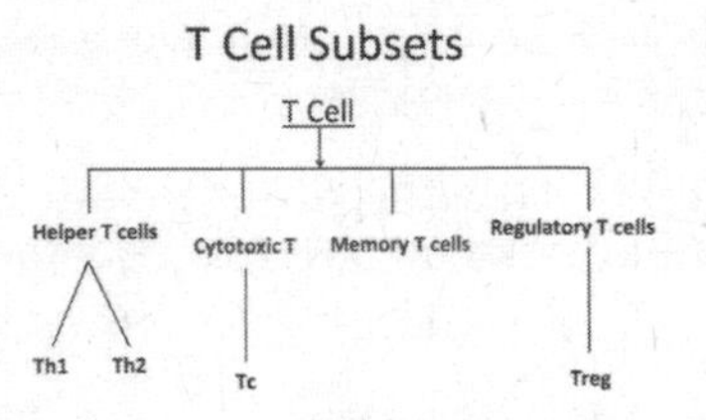

3. *the Hoskins Effect,* (page 49): refers to the propensity of the body's immune system to preferentially utilize immunological memory based on a previous infection when a second slightly different version of that foreign entity (e.g., a virus or bacterium) is encountered.
4. *plasma cells* (page 50): Plasma cells, also called plasma B cells, or effector B cells, are white blood cells that secrete large volumes of antibodies. They are transported by the blood plasma and the lymphatic system. Plasma cells originate in the bone marrow; B cells differentiate into plasma cells that produce antibody molecules closely modeled after the receptors of the precursor B cell. Once released into the blood and lymph, these antibody molecules bind to the target antigen (foreign substance) and initiate its neutralization or destruction. In *Wikipedia.* Retrieved from https://en.wikipedia.org/w/index.php?title=Plasma_cell&oldid=788883461).
5. *Goldilocks* (page 54): Goldilocks and the Three Bears is a nineteenth-century fairy tale. A little girl named Goldilocks goes for a walk in the forest and comes upon a house where she enters and finds to her delight three bowls of porridge. The first one she tastes is too hot, the next too cold but the third one is just right so she gobbles it all up. In *Wikipedia.* Retrieved from

https://en.wikipedia.org/w/index.php?title=Goldilocks_and_the_Three_Bears&oldid=786195942).

4. Mysteries of the Human Cell

1. *Adenosine triphosphate (ATP)* (page 62): is a nucleotide, which is a small molecule employed in cells as a coenzyme. It is often referred to as the "molecular unit of currency" of intracellular energy transfer. Knowles, J. R. (1980). "Enzyme-catalysed phosphoryl transfer reactions." *Annual Review of Biochemistry*, 49(1): 877–919.
2. *Endoplasmic reticulum* (page 62): is a region of the cell where proteins are made and modified in preparation for transport to the cell surface. Dodonova, S. O., Diestelkoetter-Bachert, P., Beck, R., Beck, M., et al. (2015). "A structure of the COPI coat and the role of coat proteins in membrane vesicle assembly." *Science*; 349 (6244): 195.

5. The Intelligence of Single-Celled Organisms

1. *One of the difficulties is to define life, or to differentiate between plants, organisms, and animals,* (page 74): in scientific usage, is that *organism* is a discrete and complete living thing, such as animal, plant, fungus, or microorganism while *animal* is a multicellular organism that is usually mobile, whose cells are not encased in a rigid cell wall (distinguishing it from plants and fungi) and which derives energy solely from the consumption of other organisms (distinguishing it from plants).
2. *Archaea* (page 76): are single-celled microorganisms that resemble bacteria but are different from them in certain aspects of their chemical structure, such as the composition of their cell walls. Archaea usually live in extreme, often very hot or salty environments. Despite their morphological similarity to bacteria, archaea possess genes and several metabolic pathways that are more closely related to those of eukaryotes (the group of organisms that includes plants, animals, and fungi), and enzymes notably involved in transcription and translation. They are also part of the human microbiota, found in the colon, oral cavity, and skin.
3. *mother's microbiome* (page 82): is the entirety of all genes present in the microorganisms colonizing a given host.
4. *John Tyler Bonner* (page 83): *The Social Amoebae: The Biology of the Cellular Slime Molds.* (Princeton, N.J.: Princeton University Press, 2009).
   Why Size Matters: From Bacteria to Blue Whales. (Princeton, N.J.: Princeton University Press, 2006).
   Lives of a Biologist: Adventures in a Century of Extraordinary Science. (Cambridge, Mass.: Harvard University Press, 2002).
   First Signals: The Evolution of Multicellular Development. (Princeton, N.J.: Princeton University Press, 2000).
   Sixty Years of Biology: Essays on Evolution and Development. (Princeton, N.J.: Princeton University Press, 1996).
   Life Cycles: Reflections of an Evolutionary Biologist. (Princeton, N.J.: Princeton University Press, 1992).
5. *pseudopodia* (page 84): a temporary protrusion or retractile process of the cytoplasm of a cell (such as an amoeba or a white blood cell) that functions especially as an organ of locomotion or in taking up food or other particulate matter.

6. Regeneration, Hibernation, and Metamorphosis

1. *Metamorphosing animals* (page 91): A change in the form and often habits of an animal during normal development after the embryonic stage. Metamorphosis includes, in insects,

the transformation of a maggot into an adult fly and a caterpillar into a butterfly, and in amphibians, the changing of a tadpole into a frog. https://www.merriam-webster.com/dictionary/metamorphosis

2. *instar* (page 97): Caterpillars go through five stages of growth. Each stage is called an "instar."

7. The Wisdom of the Body

1. *circadian clocks* (page 103): Noting or pertaining to rhythmic biological cycles recurring at approximately twenty-four-hour intervals.

8. Heart Transplants: Personality Transplants?

1. *cardiac natriuretic peptide* (page 120): Atrial natriuretic factor (ANF) and brain natriuretic peptide (BNP). Both hormones have potent growth-regulating properties.

9. The Brain-Mind Conundrum and the Rise of Quantum Biology

1. *rice experiment* (page 141): https://www.google.ca/search?hl=en&tbm=isch&source=hp&biw=1154&bih=728&ei=VLToWp-8IIT4jwSEp4-AAQ&q=emoto+water+rice&oq=emoto+water+rice&gs_l=img.3...1488.8373.0.9055.16.7.0.9.9.0.87.550.7.7.0....0...1ac.1.64.img..0.12.585...0.0.uKqsNGe9mMU; https://www.goodnewsnetwork.org/teacher-shows-students-how-negative-words-makes-rice-moldy/; https://yayyayskitchen.com/2017/02/02/30-days-of-love-hate-and-indifference-rice-and-water-experiment-1/.
2. *Studies of near-death experiences* (page 145): *The Handbook of Near-Death Experiences: Thirty Years of Investigation*. Janice Miner Holden, Bruce Greyson, and Debbie James, eds. (Westport, Conn.: Praeger, 2009).
   "Leaving Body and Life Behind: Out-of-Body and Near-Death Experience." Olaf Blanke, Nathan Faivre, and Sebastian Dieguez in *The Neurology of Consciousness*. 2nd edition. Steven Laureys, Olivia Gosseries, and Giulio Tononi, eds. (Cambridge, Mass.: Academic Press, 2015).
3. *extrasensory perception (ESP)* (page 146): Extrasensory perception or ESP, also called the sixth sense, refers to reception of information not gained through the recognized physical senses but with the mind. The term was adopted by Duke University psychologist J. R. Rhine in 1933.
4. *quantum physics* (page 147):

Principles of Quantum Mechanics

- Heisenberg: You can measure an electron's exact position – or you can measure the wave's momentum or speed but you cannot measure both at the same time
- A quantum object cannot be said to manifest in ordinary space-time reality until we observe it as a particle (collapse of the wave)
- All energy is transmitted in wave packets or quantums
- Unobserved quantum phenomena are radically different from observed ones
- The observation (or measurement) makes the quantum wave function collapse
- At sufficient high levels of attention energy, particles emerge from waves
- The way we observe the quantum field decide what we see, thus your belief systems determine the reality you experience

Credit: Dr. Erik Hoffmann (original text)

5. *microtubules* (page 147): https://phys.org/news/2014-01-discovery-quantum-vibrations-microtubules-corroborates.html#jCp.
6. *Anthony Hudetz* (page 148): Anthony Hudetz and Robert Pearce, eds. *Suppressing the Mind: Anesthetic Modulation of Memory and Consciousness.* Contemporary Clinical Neuroscience (New York: Humana Press, 2010). A good reference.
7. *These scientists* (page 148): International Summit on Post-Materialist Science, Spirituality, and Society : Summary Report History, Participants, Questions, Meeting, Consensus Decisions, and Representative Publications. Manifesto for a Post-Materialist Science:
   Gary E. Schwartz, PhD, host and co-organizer, University of Arizona and Canyon Ranch.
   Lisa Miller, PhD, co-organizer, Columbia University.
   Mario Beauregard, PhD, co-organizer, University of Arizona. http://opensciences.org/files/pdfs/ISPMS-Summary-Report.pdf.
8. *Stuart Alan Kauffman* (page 154): "Quantum biology on the edge of quantum chaos." G. Vattay, S. Kauffman (2014), *PLoS One.* A new approach to quantum biological systems. A highly technical but influential paper.
9. *a bridge between neuroscience and more spiritual approaches to consciousness.* (page 157): Gary E. Schwartz, PhD, is professor of psychology, medicine, neurology, psychiatry, and surgery at the University of Arizona and director of the Laboratory for Advances in Consciousness and Health. He has published more than 450 scientific papers, including three books.
   The Energy Healing Experiments: Science Reveals Our Natural Power to Heal. (New York: Atria, 2007).
   The Afterlife Experiments: Breakthrough Scientific Evidence of Life After Death. (New York: Atria, 2002).
   The Sacred Promise: How Science Is Discovering Spirit's Collaboration with Us in Our Daily Lives. (New York: Simon & Schuster, 2011). *Highly recommended.*

REFERENCES

Introduction

p. xii "Tiny brain no obstacle to French civil servant." *Reuters Science News*, July, 2007; https://www.reuters.com/article/us-brain-tiny/tiny-brain-no-obstacle-to-french-civil-servant-idUSN1930510020070720.

p. xii Dr. Lionel Feuillet of Hôpital de la Timone: Feuillet, L., Dufour, H., and Pelletier, J. (2007). "Brain of a white-collar worker." *Lancet*, 370(9583), 262.

p. xiii adults who as children: Villemure, J. G., and Rasmussen, T. H. (1993). "Functional hemispherectomy in children." *Neuropediatrics*, 24(1), 53–55; Battaglia, D., Veggiotti, P., Colosimo, C., et al. (2009). "Functional hemispherectomy in children with epilepsy and CSWS due to unilateral early brain injury including thalamus: sudden recovery of CSWS." *Epilepsy Research*, 87(2–3), 290–298; and Schramm, J., Kuczaty, S., Von Lehe, M. et al. (2012). "Pediatric functional hemispherectomy: outcome in 92 patients." *Acta Neurochirurgica*, 154(11), 2017–2028.

p. xiii adults who have had hemispherectomies: Carson, B. S., Javedan, S. P., Guarnieri, M., et al. (1996). "Hemispherectomy: A hemidecortication approach and review of 52 cases." *Journal of Neurosurgery*, 84(6), 903–911.

1. Do Genes Matter?

p. 3 "It is like you have a 100-page book": Sapolsky, Robert. *Why Zebras Don't Get Ulcers*. (New York. Henry Holt & Company, 2004). Darwin wrote: Darwin, Charles. *On the Origin of Species by Means of Natural Selection, or the Preservation of Favoured Races in the Struggle for Life*. Fellow of the Royal, Geological, Linnæan, etc. societies; Author of Journal of researches during HMS *Beagle*'s Voyage round the world. (London: John Murray, 1859).

p. 3 Following in Darwin's footsteps: Risch, N. J. (2000). "Searching for genetic determinants in the new millennium." *Nature*, 405(6788), 847–856; and Botstein, D. and Risch, N. (2003). "Discovering genotypes underlying human phenotypes: past successes for mendelian disease, future approaches for complex disease." *Nature Genetics*, 33, 228–237.

p. 4 genetic variation at a single genomic position: Dermitzakis, Emmanouil T., et al. (2015). "Population Variation and Genetic Control of Modular Chromatin Architecture in Humans." *Cell*, 162(5), 1039–1050.

p. 4 Genes teach us a crucial life lesson: Meaney, M. J. (2001). "Maternal care, gene expression, and the transmission of individual differences in stress reactivity across generations." *Annual Review of Neuroscience*, 24, 1161–1192.

p. 4 personality changes can affect body shape: Kern, Elizabeth M.A., Robinson, Detric, et al. (2016). "Correlated evolution of personality, morphology and performance." *Animal Behavior*, 117, 79. doi.org/10.1016/j.anbehav.2016.04.007.

p. 5 theoretical framework for heredity: Jose, A. M. (2020). "Heritable Epigenetic Changes Alter Transgenerational Waveforms Maintained by Cycling Stores of Information." *BioEssays*, doi.org/10.1002/bies.201900254.

p. 7 *microchimerism*: Yan, Z., Lambert N. C., Guthrie, K. A., Porter, A. J., Nelson, J. L., et al. (2005). "Male microchimerism in women without sons: Quantitative assessment and correlation with pregnancy history." *American Journal of Medicine.*, 118(8), 899–906; and Chan, W. F., Gurnot. C., Montine, T. J., Nelson, J. L., et al. (2012). "Male microchimerism in the human female brain." *PLoS One*, 7(9), e45592. doi: 10.1371/journal.pone.0045592.

p. 7 their own, their mother's, and their child's: Rowland, Katherine (2018). "We Are Multitudes." *Aeon Magazine.*

p. 8 A 1988 paper published by John Cairns: Cairns, John, Overbaugh. Julie, and Miller, Stephan (1988). "The origin of mutants." *Nature*, 335, 142–145.

p. 8 *directed mutations*: Stahl, F. W. (1988) "Bacterial genetics. A unicorn in the garden." *Nature*, 335, 112.

p. 8 genetic imprints can short-circuit evolution: Bygren, L. O., Kaati, G., and Edvinsson, S.: (2001) "Longevity determined by ancestors' overnutrition during their slow growth period." *Acta Biotheoretica*, 49, 53–59; and Bygren, Olov (2010) in Kaati, G., Bygren, L. O., Edvinsson, S. (2002). "Cardiovascular and diabetes mortality determined by nutrition during parents' and grandparents' slow growth period." *European Journal of Human Genetics*, 10, 682–688.

p. 9 *adaptive mutation*: Foster, P.L. (1999). "Mechanisms of stationary phase mutation: a decade of adaptive mutation." *Annual Review of Genetics*, 33:57–88.

p. 9 scientific articles about C. elegans: CBS C-elegans (http://cbs.umn.edu/cgc/what-c-elegans).

p. 9 80 percent of their genes: Gebauer, Juliane, Gentsch, Christoph, and Kaleta, Christoph, et al. (2016). "A Genome-Scale Database and Reconstruction of Caenorhabditis elegans Metabolism." *Cell Systems*, 2(5), 312.

p. 9 memory of famine: Jobson, M. A., Jordan, J. M., Sandrof, M. A., Baugh. L. R., et al. (2015). "Transgenerational Effects of Early Life Starvation on Growth, Reproduction and Stress Resistance in Caenorhabditis elegans." *Genetics*, 201(1), 201–212.

p. 9 famine toward the end of World War II: Tobi, Elmar W., Slieker, Roderick C., Xu, Kate M., et al. (2018). "DNA methylation as a mediator of the association between prenatal adversity and risk factors for metabolic disease in adulthood." *Science Advances*, 4(1).

p. 10 In humans, so can poverty: Pembrey, M., Saffery, R. and Bygren, L. O. "Network in Epigenetic Epidemiology. Human transgenerational responses to early-life experience: potential impact on development, health and biomedical research." *Journal of Medical Genetics.* 51, 563–72 (2015).

p. 10 *promoter regions*: Borghol, N., Suderman, M., McArdle, W., Racine, A., Hallett, M., Pembrey, M., Szyf, M., et al. (2012). "Associations with early-life socio-economic position in adult DNA methylation." *International Journal of Epidemiology*, 41(1), 62–74.

p. 11 the epigenome of an infant is sensitive: Almgren, Malin, Schlinzig, Titus, Gomez-Cabrero, David, Ekström, Tomas, J., et al. (2014). "Cesarean section and hematopoietic stem cell epigenetics in the newborn infant—implications for future health?" *American*

Journal of Obstetrics and Gynecology, 211(5), 502.e1–502.e8; and Schlinzig, T., Johansson, S., Gunnar. A., Ekström, T. J., Norman, M. (2009). "Epigenetic modulation at birth—altered DNA-methylation in white blood cells after Caesarean section." *Acta Pædiatrica*, 98, 1096–1099.

p. 11 social isolation: Conzen, Suzanne (2009). "Social isolation worsens cancer." *Institute for Genomics and Systems Biology*. Retrieved from http://www.igsb.org/news/igsb-fellow-suzanne-d.-conzen-social-isolation-in-mice-worsens-breast-cancer.

p. 11 compassionate and caring people: Slavich, G. M. and Cole, S. W. (2013). "The emerging field of human social genomics." *Clinical Psychological Science*, 1(3), 331–348.

p. 11 Kidnapping-and-Cross-Fostering Study: Zayed, Amro and Robinson, Gene E. (2012). "Understanding the Relationship Between Brain Gene Expression and Social Behavior: Lessons from the Honey Bee." *Annual Review of Genetics*, 46, 591–615.

p. 12 if a male zebra finch heard another male zebra finch: Clayton, D. F. and London, S. E. (2014). "Advancing avian behavioral neuroendocrinology through genomics." *Frontiers in Neuroendocrinology*, 35(1), 58–71.

p. 13 difference between a calm and an anxious rat: Weaver, I.C.G., Cervoni, N., Champagne, F. A., Meaney, M. J., et al. (2004). "Epigenetic programming by maternal behavior." *Nature Neuroscience*, 7, 847–854.

p. 13 distress responses in the infants: Roth, T. L., Lubin, F. D., Funk, A. J., and Sweatt, J. D. (2009). "Lasting epigenetic influence of early-life adversity on the BDNF gene." *Biological Psychiatry*, 65(9), 760–769.

p. 13 The maltreated infant rats: Roth, Tania L. and Sweatt, J. David (2011). "Epigenetic mechanisms and environmental shaping of the brain during sensitive periods of development." *Journal of Child Psychology and Psychiatry*, 52(4), 398–408; and Roth, Tania L. and Sweatt, J. David (2011). "Epigenetic marking of the BDNF gene by early-life adverse experiences." *Hormones and Behavior,* Special Issue: Behavioral Epigenetics, 59(3), 315–320.

p. 13 childhood maltreatment: Bremner, J. D. (2003). "Long-term effects of childhood abuse on brain and neurobiology." *Child and Adolescent Psychiatric Clinics of North America*, 12, 271–292; Heim, C. and Nemeroff, C. B. (2003). "The role of childhood trauma in the neurobiology of mood and anxiety disorders: preclinical and clinical studies." *Biological Psychiatry*, 49, 1023–1039; Kaufman, J., Plotsky, P. M., Nemeroff, C. B., and Charney, D. S. (2000). "Effects of early adverse experiences on brain structure and function: clinical implications." *Biological Psychiatry*, 48, 778–790; and Schore, A. N. (2002). "Dysregulation of the right brain: a fundamental mechanism of traumatic attachment and the psychopathogenesis of posttraumatic stress disorder." *Australian and New Zealand Journal of Psychiatry*, 36, 9–30.

p. 14 positive short-term behavioral response in her child: Sullivan, Regina. NYU Langone Medical Center/New York University School of Medicine. "Mother's soothing presence makes pain go away, changes gene activity in infant brain." *ScienceDaily*, November 18, 2014. www.sciencedaily.com/releases/2014/11/141118125432.htm.

p. 14 distinct smell experience of parents: Dias, B. G., Maddox, S. A., Klengel, T., and Ressler, K. J. (2015). "Epigenetic mechanisms underlying learning and the inheritance of learned behaviors." *Trends in Neurosciences*, 38(2), 96–107.

p. 15 Molecules in the sperm called microRNAs: Morgan, Christopher P. and Bale, Tracy L. (2011). "Early prenatal stress epigenetically programs dysmasculinization in second-generation offspring via the paternal lineage." *Journal of Neuroscience*, 17, 31(33), 11748–11755.

p. 15 effect of stress in four generations of rats: Yao, Y., Robinson, A. M., Metz, G. A., et al. (2014). "Ancestral exposure to stress epigenetically programs preterm birth risk and adverse maternal and newborn outcomes." *BMC Medicine*, 12(1), 121.

p. 15 exposed male mice to low-dose nicotine: McCarthy, Deirdre M., Morgan, Thomas J., Bhide, Pradeep G., et al. (2018). "Nicotine exposure of male mice produces behavioral impairment in multiple generations of descendants." *PLoS Biology* 16(10), e2006497.

p. 16 girls whose maternal grandmother smoked: Golding, Jean, Ellis, Genette, Pembrey, Marcus, et al. (2017). "Grand-maternal smoking in pregnancy and grandchild's autistic traits and diagnosed autism." *Scientific Reports* 7, article number 46179.

p. 16 the sperm of males who smoked or ingested marijuana: Schrott, R., Acharya, K., Murphy, S. K., et al. (2020). "Cannabis use is associated with potentially heritable widespread changes in autism candidate gene DLGAP2 DNA methylation in sperm." *Epigenetics*, 15(1–2), 161–173.

p. 16 Researchers from Mount Sinai School of Medicine: Dietz, D. M., LaPlant, Q., Watts, E. L., Hodes, G. E., Russo, S. J., Feng, J., et al. (2011). "Paternal transmission of stress-induced pathologies." *Biological Psychiatry*, 70, 408–414.

p. 17 Transgenerational Trauma: Freud, A., and Burlingham, D. *War and Children*. (New York: International University Press, 1942).

p. 18 a mother's and a child's psychic borders: Mahler, M. S. (1968.) "On Human Symbiosis and the Vicissitudes of Individuation". *Journal of the American Psychoanalytic Association*. doi.org/10.1177/000306516701500401.

p. 18 "historical trauma": Brave Heart, M.Y.H., Chase, J., Elkins, J., and Altschul, D. B. (2011). "Historical trauma among indigenous peoples of the Americas: Concepts, research, and clinical considerations." *Journal of Psychoactive Drugs*, 43(4), 282–290.

p. 18 "walking on eggshells" around the PTSD parent: Volkan, Vamik, D. (1998). "Transgenerational Transmissions and Chosen Traumas." Opening Address, XIII International Congress. International Association of Group Psychotherapy. http://www.vamikvolkan.com/Transgenerational-Transmissions-and-Chosen-Traumas.php.

p. 19 studies on the children of Holocaust survivors: Barocas, H. A. and Barocas, C. B. (1979). "Wounds of the fathers: The next generation of Holocaust victims." *International Review of Psycho-Analysis*; Freyberg, J. T. (1980). "Difficulties in separation-individuation as experienced by holocaust survivors." *American Journal of Orthopsychiatry*, 50(1), 87–95; and Fogelman, E. and Savran, B. (1980). "Brief group therapy with offspring of Holocaust survivors: Leaders' reactions." *American Journal of Orthopsychiatry*, 50(1), 96.

p. 19 neuroendocrine (hormonal) abnormalities: Sorscher, N. and Cohen, L. J. (1997). "Trauma in children of Holocaust survivors: Transgenerational effects." *American Journal of Orthopsychiatry*, 67(3), 493; Yehuda, R., Halligan, S. L., and Grossman, R. (2001). "Childhood trauma and risk for PTSD: relationship to intergenerational effects of trauma, parental PTSD, and cortisol excretion." *Development and Psychopathology*, 13(3), 733–753; and Yehuda, Rachel, Daskalakis, Nikolaos P., Binder, Elisabeth B., et al. (2016). "Holocaust Exposure Induced Intergenerational Effects on FKBP5 Methylation." *Biological Psychiatry*, 80(5), 372, doi: 10.1016/j.biopsych.2015.08.005.

p. 19 persons with post-traumatic stress disorder: Yehuda, R., Daskalakis, N. P., Lehrner, A., Desarnaud, F., Bader, H. N., Makotkine, I., et al. (2014). "Influences of maternal and paternal PTSD on epigenetic regulation of the glucocorticoid receptor gene in Holocaust survivor offspring." *American Journal of Psychiatry*, 171(8), 872–8010.1176/appi.ajp.2014.13121571.

p. 19 study on thirty-eight women who were pregnant on 9/11: Yehuda, R., et al. (2005). "Transgenerational Effects of Posttraumatic Stress Disorder in Babies of Mothers Exposed to the World Trade Center Attacks during Pregnancy." *Journal of Clinical Endocrinology & Metabolism*, doi: 10.1210/jc.2005–0550.

p. 19 mothers who were in their second or third trimester: Sarapas, C., et al. (2011). "Genetic markers for PTSD risk and resilience among survivors of the World Trade Center attacks." *Disease Markers*, 30(2-3), 101–110.

p. 19 artificially separating male mice from their mothers: Gapp, Katharina, Bohacek, Johannes, Mansuy, Isabelle M., et al. (2016). "Potential of Environmental Enrichment to Prevent Transgenerational Effects of Paternal Trauma." *Neuropsychopharmacology*; and Gapp, Katharina, Jawaid, Ali, Sarkies, Peter, Mansuy, Isabelle M., et al. (2014). "Implication of sperm RNAs in transgenerational inheritance of the effects of early trauma in mice." *Nature Neuroscience*, 17, 667–669, doi:10.1038/nn.3695.

p. 19 Multiple rodent and nonhuman primate studies: Branchi, I., Francia, N., and Alleva, E. (2004). "Epigenetic control of neurobehavioral plasticity: the role of neurotrophins." *Behavioral Pharmacology*, 15, 353–362; and Roth, T. L., Lubin, F. D., Funk, A. J., and Sweatt, J. D. (2009). "Lasting epigenetic influence of early-life adversity on the BDNF gene." *Biological Psychiatry*, 65(9), 760–769.

p. 19 lasting changes in neural function and behavior: Korosi, A. and Baram, T. Z. (2009). "The pathways from mother's love to baby's future." *Frontiers in Behavioral Neuroscience*. Epub ahead of print September 24, 2009; Pryce, C. R. and Feldon, J. (2003). "Long-term neurobehavioral impact of postnatal environment in rats: manipulations, effects and mediating mechanisms." *Neuroscience and Biobehavioral Reviews*; 27:57–71; and Sanchez, M. M. (2006). "The impact of early adverse care on HPA axis development: Nonhuman primate models." *Hormones and Behavior*. 50:623–631

p. 20 The offspring for two successive generations: Gapp, Katharina, Bohacek, Johannes, Mansuy, Isabelle M., et al. (2016). "Potential of Environmental Enrichment to Prevent Transgenerational Effects of Paternal Trauma." *Neuropsychopharmacology*, doi: 10.1038/npp.2016.87; and Gapp, Katharina, Jawaid, Ali, Sarkies, Peter, Mansuy, Isabelle M., et al. (2014). "Implication of sperm RNAs in transgenerational inheritance of the effects of early trauma in mice." *Nature Neuroscience*, 17, 667–669, doi:10.1038/nn.3695.

p. 20 the critical role that fathers play: Siklenka, Keith, Erkek, Serap, Kimmins, Sarah, et al. (2015). "Disruption of histone methylation in developing sperm impairs offspring health transgenerationally." *Science*, 350(6261).

p. 20 small noncoding RNAs called *micro RNAs (miRNAs)*: Ebert, M. S. and Sharp, P. A. (2012) "Roles for microRNAs in conferring robustness to biological processes." *Cell*, 149, 515–524; and Biggar, K. K. and Storey, K. B. (2011) "The emerging roles of microRNAs in the molecular responses of metabolic rate depression." *Journal of Molecular Cell Biology*, 3, 167–175.

p. 20 newly discovered type of gene regulator: Morgan, Christopher P. and Bale, Tracy L. (2011). "Early prenatal stress epigenetically programs dysmasculinization in second-generation offspring via the paternal lineage." *Journal of Neuroscience*, 17, 31(33), 11748–11755.

p. 21 In a Tufts University School of Medicine study: Gapp, Katharina, Jawaid, Ali, Sarkies, Peter, Mansuy, Isabelle M, et al. (2014). "Implication of sperm RNAs in transgenerational inheritance of the effects of early trauma in mice." *Nature Neuroscience*, 17, 667–669, doi:10.1038/nn.3695.

p. 21 defending the body from invasions by viruses: Sadanand, Fulzele, Bikash, Sahay, Carlos, M. Isales, et al. (2020). "COVID-19 Virulence in Aged Patients Might Be

Impacted by the Host Cellular MicroRNAs Abundance/Profile." *Aging and Disease*, 11(3), 509–522.

p. 22 sperm can be vulnerable to environmental factors: Morgan, Christopher P. and Bale, Tracy L. (2011). "Early prenatal stress epigenetically programs dysmasculinization in second-generation offspring via the paternal lineage." *Journal of Neuroscience*, 17, 31(33), 11748–11755.

p. 22 evidence is accumulating: Donkin, I. and Barrès, R. (2018). "Sperm epigenetics and influence of environmental factors." *Molecular Metabolism*, 14, 1–11.

p. 22 A related study at the University of Massachusetts Medical School: Conine, C. C., Sun, F., Rando, O. J., et al. (2018). "Small RNAs gained during epididymal transit of sperm are essential for embryonic development in mice." *Developmental Cell*, 46(4), 470–480.

p. 22 newly discovered form of communication among cells: Andaloussi, S. E., Mäger, I., Breakefield, X. O., and Wood, M. J. (2013). "Extracellular vesicles: biology and emerging therapeutic opportunities." *Nature Reviews Drug Discovery*, 12(5), 347–357.

p. 22 "It is high time public health researchers": Griffin, Matthew (August 29, 2016) "Researchers Find Evidence That Ancestors Memories Are Passed Down in DNA." *Enhanced Humans and Biotech*. https://www.311institute.com/researchers-find-evidence-that-ancestors-memories-are-passed-down-in-dna/.

p. 23 epigenetic transformation by way of self-regulation: Kaliman, Perla, Alvarez-Lopez, Maria Jesus, Davidson, Richard J., et al. (2014). "Rapid changes in histone deacetylases and inflammatory gene expression in expert meditators." *Psychoneuroendocrinology*, 40, 96–107.

p. 23 eighty-four hotel maids in New York into two groups: Crum, Alia J. and Langer, Ellen J. (2007). "Mind-Set Matters: Exercise and the Placebo Effect." *Psychological Science*, 18(2), 165–171.

p. 24 listening to Mozart's Violin Concerto: Kanduri, Chakravarthi, Raijas, Pirre, Järvelä, Irma (2015). "The effect of listening to music on human transcriptome." *PeerJ*, 23, e830, doi: 10.7717/peerj.830.

p. 24 aspirin and hormonal replacement therapy: Noreen, Faiza, Röösli, Martin, Gaj, Pawel, et al. (2014). "Modulation of Age- and Cancer-Associated DNA Methylation Change in the Healthy Colon by Aspirin and Lifestyle." *Journal of the National Cancer Institute*, 07/2014, 106(7).

p. 24 we are architects of our own lives: Cole, Steve W. (2009). "Social Regulation of Human Gene Expression." *Current Directions in Psychological Science*, 18(3), 132–137.

p. 25 association between optimism and physical health: Rasmussen, H. N., Scheier, M. F., and Greenhouse, J. B. (2009). "Optimism and physical health: A meta-analytic review." *Annals of Behavioral Medicine*, 37(3), 239–256.

p. 25 People who feel enthusiastic, hopeful and cheerful: Hittner, E. F., Stephens, J. E., Turiano, N. A., Gerstorf, D., Lachman, M. E., and Haase, C. M. (2020). "Positive Affect Is Associated With Less Memory Decline: Evidence From a 9-Year Longitudinal Study." *Psychological Science*, 0956797620953883.

p. 25 bipolar illness: Ossola, Paolo, Garrett, Neil, Sharot, Tali, and Marchesi, Carlo (2020). "Belief updating in bipolar disorder predicts time of recurrence." *eLife*, 9.

p. 25 people suffering from depression: Zenger, M., Glaesmer, H., Höckel, M., and Hinz, A. (2011). "Pessimism predicts anxiety, depression, and quality of life in female cancer patients." *Japanese Journal of Clinical Oncology*, 41(1), 87–94.

p. 26 epigenetic hypothesis for environmental contributions: Costa, E., Chen, Y., Dong, E., Grayson, D. R., Kundakovic, and M., Guidotti, A. (2009). "GABAergic promoter hypermethylation as a model to study the neurochemistry of schizophrenia vulnerability." *Expert Review of Neurotherapeutics*, 9, 87– 98, McGowan, P. O. and Szyf, M. (2010). "The epigenetics of social adversity in early life: Implications for mental health outcomes." *Neurobiology of Disease*, 39(1), 66–72; McGowan, P.O., Meaney, M.J., and Szyf, M. (2008). "Diet and the epigenetic (re)programming of phenotypic differences in behavior." *Brain Research*, 1237, 12–24; and Roth, T. L., Lubin, F. D., Sodhi, M., and Kleinman, J. E. (2009). "Epigenetic mechanisms in schizophrenia." *Biochimica et Biophysica Acta (BBA)*—General Subjects, 1790:869–877.

p. 26 before their offspring is even conceived: Lane, Michelle, Robker, Rebecca L., and Robertson, Sarah A. (2014). "Parenting from before conception." *Science*, August 15, 2014, 756–760.

p. 26 not only of their children, but that of their grandchildren: Roth T. L., Lubin, F. D., Funk, A. J. and Sweatt, J. D. (2009). "Lasting epigenetic influence of early-life adversity on the BDNF gene." *Biological Psychiatry*, 65(9), 760–769; and Tsankova, N., Renthal, W., Kumar, A., Nestler, E. J. (2007). "Epigenetic regulation in psychiatric disorders." *Nature Reviews Neuroscience*, 8, 355–367.

p. 26 in many ways, several generations further back: Yehuda, R. and Lehrner, A. (2018). "Intergenerational transmission of trauma effects: putative role of epigenetic mechanisms." *World Psychiatry*, 17(3), 243–257.

2. The Brain: How It Remembers What It Remembers

p. 29 autonomic nervous system: Schmidt, A. and Thews, G. "Autonomic Nervous System" in Janig, W., *Human Physiology* (2nd ed.). (New York: Springer-Verlag, 1989), 333–370.

p. 30 A team of researchers: Eyal, G., Verhoog, M. B., Segev, I., et al. (2016). "Unique membrane properties and enhanced signal processing in human neocortical neurons." *eLife*, 5, e16553.

p. 31 *glial cells*: Fields, R. Douglas. *The Other Brain*. (New York: Simon & Schuster, 2011).

p. 31 Glial cells can affect the functioning of neurons: Jab, Ferris (2012). "Know Your Neurons: What Is the Ratio of Glia to Neurons in the Brain?" *Scientific American*.

p. 31 glial cells control communication between neurons: Fields, R. Douglas (2010). "Visualizing Calcium Signaling in Astrocytes." *Science Signaling*, 3(147).

p. 32 experience builds memories in synapses: Christianson, S. A. (1992). "Emotional stress and eyewitness memory: a critical review." *Psychological Bulletin*, 112(2), 284–309.

p. 32 learning induces modification of synapses: Myhrer, T. (2003). "Neurotransmitter systems involved in learning and memory in the rat: a meta-analysis based on studies of four behavioral tasks." *Brain Research Brain Research Reviews*, 41(2–3), 268–87.

p. 32 storage of memories in an ever-changing plastic brain: Gonzalez, Walter G., Zhang, Hanwen, Harutyunyan, Anna, and Lois, Carlos (2019). "Persistence of neuronal representations through time and damage in the hippocampus." *Science*, 365(6455), 821–825.

p. 32 established fact and "generally accepted.": Bruel-Jungerman, E., Davis, S., and Laroche, S. (2007). "Brain plasticity mechanisms and memory: a party of four." *The Neuroscientist*, 13, 492–505. doi: 10.1177/1073858407302725.

p. 32 According to this reigning neuroscientific view: Gaidos, Susan (2013). "Memories lost and found: Drugs that help mice remember reveal role for epigenetics in recall." *Science News*.

p. 32 highly problematic notion: Delgado-García, J. M. and Gruart, A. (2004). "Neural plasticity and regeneration: myths and expectations," in *Brain Damage and Repair: From Molecular Research to Clinical Therapy*, T. Herdegen and J. M. Delgado-García, eds. (Dordrecht, Netherlands: Springer, 2004), 259–273; Delgado-García, J. M. (2015). "Cajal and the conceptual weakness of neural sciences." *Frontiers in Neuroanatomy*, 9, 128. doi: 10.3389/fnana.2015.00128.

p. 32 Retrieval of emotionally charged memories: Wagner, A. D. and Davachi, L. (2001). "Cognitive neuroscience: forgetting of things past." *Current Biology*, 11, R964–967; and Buchanan, Tony W. (2007). "Retrieval of Emotional Memories." *Psychological Bulletin*, 61–779. doi: 10.1037/0033–2909.133.5.761.

p. 33 basic building block of memory is the synapse: Kandel, Eric R. (2002). "The Molecular Biology of Memory Storage: A Dialog Between Genes and Synapses." *Science*, 21(5), 567.

p. 33 major conceptual shifts must take place: Kandel, E. R. *In Search of Memory: The Emergence of a New Science of Mind.* (New York: W. W. Norton, 2007).

p. 33 radical modification of the standard model of memory storage: Fusi, S. and Abbott, L. F. (2007). "Limits on the memory storage capacity of bounded synapses." *Nature Neuroscience*, 10(4), 485; and Firestein, S. *Ignorance: How It Drives Science.* (New York: Oxford University Press, 2012).

p. 34 memories are stored inside of neurons: Chen, Shanping, Cai, Diancai, Glanzman, David L., et al. (2014). "Reinstatement of long-term memory following erasure of its behavioral and synaptic expression in *Aplysia*." *eLife*, 3, e03896.

p. 34 In the same vein, researchers at the University of Pennsylvania: Shelley L. Berger (May 31, 2017). "Metabolic enzyme fuels molecular machinery of memory." *Penn Medicine News*; Penn study finds epigenetics key to laying down spatial memories in mouse brain, providing possible new neurological medications. https://www.pennmedicine.org/news/news-releases/2017/may/metabolic-enzyme-fuels-molecular-machinery-of-memory.

p. 34 ACSS2 works directly within the nucleus of neurons: Mews, Philipp, Donahue, Greg, Berger, Shelley L., et al. (2017). "Acetyl-CoA synthetase regulates histone acetylation and hippocampal memory." *Nature*; and Gräff, J. and L. H. Tsai (2013). "Histone acetylation: molecular mnemonics on the chromatin." *Nature Reviews Neuroscience*, 14, 97–111.

p. 34 when new memories are being established: Myhrer, T. (2003). "Neurotransmitter systems involved in learning and memory in the rat: a meta-analysis based on studies of four behavioral tasks." *Brain Research Brain Research Reviews*, 41(2–3), 268–87.

p. 34 researchers in the laboratory of Carlos Lois: Gonzalez, Walter G., Zhang, Hanwen, Harutyunyan, Anna, Lois, Carlos (2019). "Persistence of neuronal representations through time and damage in the hippocampus." *Science*, 365(6455), 821–825.

p. 34 synapse is an ill fit: Trettenbrein, P. C. (2016). "The demise of the synapse as the locus of memory: A looming paradigm shift?" *Frontiers in Systems Neuroscience*, 10, 88.

p. 34 To enable cognition and the storage of memories: "Scientists advance search for memory's molecular roots: Architecture of the cytoskeleton in neurons." *ScienceDaily* (August 26, 2019). Retrieved November 22, 2020 from www.sciencedaily.com/releases/2019/08/190826150658.htm.

p. 35 researchers in Germany: Gidon, A., Zolnik, T. A., Larkum, M. E., et al. (2020). "Dendritic action potentials and computation in human layer 2/3 cortical neurons." *Science*, 367(6473), 83–87.

p. 35 "a single neuron could be a complex computational device": Gary Marcus in Cepelewicz, Jordana (January 14, 2020) "Hidden Computational Power Found in the Arms of Neurons." *Quanta Magazine*. Retrieved from https://www.quantamagazine.org/neural-dendrites-reveal-their-computational-power-20200114/.

p. 35 Further research into how neurons perform their tasks: Poulopoulos, A. and Macklis, J. D., et al. (2019). "Subcellular transcriptomes and proteomes of developing axon projections in the cerebral cortex." *Nature*, 565(7739), 356–360.

p. 36 Neurons are elegant cells: Fields, R. Douglas (2013). "Human Brain Cells Make Mice Smart." *Scientific American*.

p. 36 long considered of little significance: Pinto-Duarte, A., Roberts, A. J., Ouyang, K., and Sejnowski, T. J. (2019). "Impairments in remote memory caused by the lack of Type 2 IP3 receptors." *Glia*. 10.1002/glia.23679.

p. 36 disabling the release of gliotransmitters: Lee, H. S., Ghetti, A., Pinto-Duarte, A., Wang, X., Dziewczapolski, G., Galimi, F., and Sejnowski, T. J. (2014). "Astrocytes contribute to gamma oscillations and recognition memory." *Proceedings of the National Academy of Sciences*, 111(32), E3343–E3352.

p. 37 using neurotransmitters: Han, X., Chen, M., Silva, A. J., et al. (2013). "Forebrain engraftment by human glial progenitor cells enhances synaptic plasticity and learning in adult mice." *Cell Stem Cell*, 12(3), 342–353.

p. 37 astroglia nets: Colombo, J. A. and Reisin, H. D. (2004) "Interlaminar astroglia of the cerebral cortex: A marker of the primate brain." *Brain Research*, 1006, 126–131.

p. 37 researchers from the National University of Ireland, Galway: Holleran, Laurena, Kelly, Sinead, Donoho, Gary, et al. (2020). "The Relationship Between White Matter Microstructure and General Cognitive Ability in Patients with Schizophrenia and Healthy Participants in the ENIGMA Consortium." *American Journal of Psychiatry*, appi.ajp.2019.1.

p. 37 When considering how the brain works: Robert Malenka in Fields, R. Douglas (2013). "Human Brain Cells Make Mice Smart." *Scientific American*; and Fields, Douglas (2013). "'Brainy' Mice with Human Brain Cells: Chimeras of Mice and Men." *BrainFacts*.

p. 37 spinal cord might learn motor skills: Vahdat, Shahabeddin, Lungu, Doyon, Ovidiu, Julien, et al. (2015). "Simultaneous Brain–Cervical Cord fMRI Reveals Intrinsic Spinal Cord Plasticity during Motor Sequence Learning." *PLoS*. http://dx.doi.org/10.1371/journal.pbio.1002186.

p. 38 *enteric nervous system*: Gershon, M. D. (2020). "The thoughtful bowel." *Acta Physiologica*, 228(1), e13331.

p. 38 *The Second Brain*: Gershon, Michael D. *The Second Brain*. (New York: HarperCollins: 1998).

p. 38 "smartness" of the ENS: Schemann, M., Frieling, T., and Enck, P. (2020). "To learn, to remember, to forget—How smart is the gut?" *Acta Physiologica*, 228(1), e13296.

p. 38 badly in need of an upgrade: Trettenbrein, P. (2016). "The Demise of the Synapse as the Locus of Memory: A Looming Paradigm Shift?" *Frontiers in Systems Neuroscience*, 10. doi: 10.3389/fnsys.2016.00088.

p. 39 Crows and ravens: Stacho, Martin, Herold, Christina, Güntürkün, Onur, et al. (2020). "A cortex-like canonical circuit in the avian forebrain." *Science*.

p. 39 Octopuses lack a central brain: Beblo, Julienne. (Oct. 16, 2018). "Are You Smarter Than an Octopus?" *National Marine Sanctuary Foundation*. https://marinesanctuary.org/blog/are-you-smarter-than-an-octopus/.

p. 39 Radical removal of half of the brain: McClelland III, S. and Maxwell, R. E. (2007). "Hemispherectomy for intractable epilepsy in adults: The first reported series." *Annals of Neurology*, 61(4), 372–376.

p. 39 epilepsy in children: Villemure, J. G. and Rasmussen, T. H. (1993). Functional hemispherectomy in children. *Neuropediatrics*, 24(01), 53–55; Battaglia, D., Veggiotti, P., Lettori, D., Tamburrini, G., Tartaglione, T., Graziano, A., Colosimo, C. (2009). "Functional hemispherectomy in children with epilepsy and CSWS due to unilateral early brain injury including thalamus: sudden recovery of CSWS." *Epilepsy Research*, 87(2–3), 290–298; and Schramm, J., Kuczaty, S., Sassen, R., Elger, C. E., and Von Lehe, M. (2012). "Pediatric functional hemispherectomy: outcome in 92 patients." *Acta Neurochirurgica*, 154(11), 2017–2028.

p. 39 team at Johns Hopkins University in Baltimore: Carson, B. S., Javedan, S. P., Guarnieri, M., et al. (1996). "Hemispherectomy: a hemidecortication approach and review of 52 cases." *Journal of Neurosurgery*, 84(6), 903–911.

p. 39 "awed by the apparent retention of memory": Vining, E. P., Freeman, J. M., Pillas, D. J., Uematsu, S., Zuckerberg, A., et al. (1997). "Why would you remove half a brain? The outcome of 58 children after hemispherectomy—the Johns Hopkins experience: 1968 to 1996." *Pediatrics*, 100(2), 163–171.

p. 40 case from China: Feng Yu, Qing-jun Jiang, Xi-yan Sun, Rong-wei Zhang; (2015). "A new case of complete primary cerebellar agenesis: clinical and imaging findings in a living patient." *Brain*, 138(6), e353.

p. 40 ten-year-old German girl: Muckli, Lars (2009). "Scientists reveal secret of girl with 'all seeing eye'." *University of Glasgow*. http://www.gla.ac.uk/news/archiveofnews/2009/july/headline_125704_en.html.

p. 40 series of six hundred cases: Lewin, R. (1980). "Is your brain really necessary?" *Science* 210(4475), 1232–1234 10.1126/science.6107993; and Lorber J. (1978). "Is Your Brain Really Necessary?" *Archives of Disease in Childhood*, 53(10), 834–835.

p. 40 instead of the typical 4.5 cm thickness of brain tissue: Lorber, J. (1978). "Is Your Brain Really Necessary?" *Archives of Disease in Childhood*, 53(10), 834–835.

p. 41 In July 2007, a forty-four-year-old Frenchman: Feuillet, L.; Dufour, H.; Pelletier, J. (2007). "Brain of a white-collar worker." *Lancet*, 370(9583), 262.

p. 41 "What I find amazing to this day": "Tiny brain no obstacle to French civil servant." *Reuters.*

p. 41 Neuroscientists explain the near normal behavior: Wojtowicz, J. M. (2011). "Adult neurogenesis. From circuits to models." *Behavioral Brain Research*. [Epub ahead of print]. 10.1016/j.bbr.2011.08.013.

p. 42 Logic dictates that there must be a limit: Majorek, M. B. (2012). "Does the brain cause conscious experience?" *Journal of Consciousness Studies*, 19, 121–144.

p. 42 Jason Padgett: Markovich, Matt (2015). "Blow to the head turns Tacoma man into a genius." *Komono News*, https://komonews.com/news/local/knocked-out-in-bar-fight-man-wakes-up-as-a-genius-11-21-2015.

p. 43 "recruiting" another part of the brain: Treffert, Darald *Extraordinary People: Understanding the Savant Syndrome.* (Lincoln, Neb.: iUniverse, 2006), originally published by Ballantine and Treffert, Darold (2015). "Genetic Memory: How We Know Things We Never Learned." *Scientific American.*

p. 43 the brain comes loaded: Gazzaniga, Michæl S. *The Mind's Past.* (Berkeley: University of California Press, 2000).

p. 44 critical role for polyunsaturated fatty acids: Zamroziewicz, Marta K.; Paul, Erick J., Aron, K., et al. (2017). "Determinants of fluid intelligence in healthy aging: Omega-3 polyunsaturated fatty acid status and frontoparietal cortex structure." *Nutritional Neuroscience*, 1–10; and Gu, Y., Vorburger, R. S., Brickman, A. M., et al. (2016). "White matter integrity as a mediator in the relationship between dietary nutrients and cognition in the elderly." *Annals of Neurology*, 79(6), 1014–1025.

p. 45 consumption of extra-virgin olive oil: Lauretti, E., Iuliano, L., and Praticò, D. (2017). "Extra-virgin olive oil ameliorates cognition and neuropathology of the 3xTg mice: role of autophagy." *Annals of Clinical and Translational Neurology*, 4(8), 564–574.

p. 45 study recently published by Jason Steffener: Steffener, Jason, Habeck, Christian, Stern, Yaakov, et al. (2016). "Differences between chronological and brain age are related to education and self-reported physical activity." *Neurobiology of Aging*, 40, 138. doi.

p. 45 Green spaces are good for gray matter: Tilley, Sara, Neale, Chris, Patuano, Agnès, and Cinderby, Steve (2017). Older People's Experiences of Mobility and Mood in an Urban Environment: A Mixed Methods Approach Using Electroencephalography (EEG) and Interviews. *International Journal of Environmental Research and Public Health*, 14(2), 151. Read more at https://medicalxpress.com/news/2017-04-green-spaces-good-grey.html#jCp

p. 46 astrocyte networks may serve as nonneuronal channels: Colombo, J. A. and Reisin, H. D. (2004), "Interlaminar astroglia of the cerebral cortex: a marker of the primate brain." *Brain Research*, 1006, 126–131.

p. 46 functional assemblies: Wixted, J. T., Squire, L. R., Steinmetz, P. N., et al. (2014). "Sparse and distributed coding of episodic memory in neurons of the human hippocampus." *Proceedings of the National Academy of Sciences*, 111(26), 9621–9626. 1.

3. The Immune System: This Doctor Makes House Calls

p. 50 Peter Cockerill, from the University of Birmingham: Bevington, S. L., Cauchy, P., Cockerill, P. N., et al. (2016). "Inducible chromatin priming is associated with the establishment of immunological memory in T cells." *The EMBO Journal*, 35(5), 515–535.

p. 50 Thomas Dörner, from the University of Berlin: Dörner, T. and Radbruch, A. (2007). "Antibodies and B cell memory in viral immunity." *Immunity*, 27(3), 384–392.

p. 50 NK cells in the liver: Stary, V., Pandey, R. V., Stary, G., et al. (2020). "A discrete subset of epigenetically primed human NK cells mediates antigen-specific immune responses." *Science Immunology*, 5(52).

p. 50 Australian scientists have discovered: Suan, D., Nguyen, A., Kaplan, W., et al. (2015). "T follicular helper cells have distinct modes of migration and molecular signatures in naive and memory immune responses." *Immunity*, 42(4), 704–718.

p. 51 This is an important finding: Moran, I., Nguyen, A., Munier, C.M.L., et al. (2018). "Memory B cells are reactivated in subcapsular proliferative foci of lymph nodes." *Nature Communications*, 9(1), 1–14.

p. 51 speaking of lymph nodes, Jochen Hühn: Cording, S., Hühn, J., Pabst, O., et al. (2013). "The intestinal micro-environment imprints stromal cells to promote efficient Treg induction in gut-draining lymph nodes." *Mucosal Immunology*, 7(2), 359–368.

p. 51 after infection with the parasitic disease leishmaniasis: Glennie, N. D., Yeramilli, V. A., Beiting, D. P., Volk, S. W., Weaver, C. T., and Scott, P. (2015). "Skin-resident memory CD4+ T cells enhance protection against Leishmania major infection." *Journal of Experimental Medicine*, jem-20142101.

p. 51 When activated by a pathogen or stress: Bachiller, S., Jiménez-Ferrer, I., and Boza-Serrano, A., et al. (2018). "Microglia in neurological diseases: a road map to brain-disease dependent-inflammatory response." *Frontiers in Cellular Neuroscience*, 12, 488.

p. 52 discovery of meningeal lymphatic vessels: Derecki, N. C., Cardani, A. N., Yang, C. H., Quinnies, K. M., Crihfield, A., Lynch, K. R., and Kipnis, J. (2010). "Regulation of learning and memory by meningeal immunity: a key role for IL-4." *Journal of Experimental Medicine*, 207(5), 1067–1080

p. 52 viral infection during pregnancy: Warre-Cornish, K., Perfect, L., Srivastav, Deepak P., McAlonan, G., et al. (2020). "Interferon-γ signaling in human iPSC–derived neurons recapitulate neurodevelopmental disorder phenotypes." *Science Advances*, 6(34), eaay9506.

p. 53 development of autism spectrum disorder: Filiano, A. J., Gadani, S. P., and Kipnis, J. (2015). "Interactions of innate and adaptive immunity in brain development and function." *Brain Research*, 1617, 18–27.

p. 53 research of mania: Dickerson, F., Adamos, M., Yolken, R. H., et al. (2018). "Adjunctive probiotic microorganisms to prevent rehospitalization in patients with acute mania: a randomized controlled trial." *Bipolar Disorders*, 20(7), 614–621.

p. 54 postpartum depression: Ohio State University. (November 6, 2018). "Immune system and postpartum depression linked? Research in rats shows inflammation in brain region after stress during pregnancy." *ScienceDaily*.

p. 55 mind-body practices like meditation: Fredrickson, B. L., Grewen, K. M., Cole, S. W., et al. (2013). "A functional genomic perspective on human well-being." *Proceedings of the National Academy of Sciences*, 110(33), 13684–13689.

p. 55 higher antibody production: Davidson, R. J., Kabat-Zinn, J., Sheridan, J. F., et al. (2003). "Alterations in brain and immune function produced by mindfulness meditation." *Psychosomatic Medicine*, 65(4), 564–570.

p. 55 negative inflammatory activity: Pace, T. W., Cole, S. P., Raison, C. L., et al. (2009). "Effect of compassion meditation on neuroendocrine, innate immune and behavioral responses to psychosocial stress." *Psychoneuroendocrinology*, 34(1), 87–98.

p. 55 positive antiviral response: Morgan, N., Irwin, M. R., Chung, M., and Wang, C. (2014). "The effects of mind-body therapies on the immune system: meta-analysis." *PLoS One*, 9(7), e100903.

p. 55 function of specific strains of immune cells: Fang, C. Y., Reibel, D. K., Longacre, M. L., Douglas, S. D., et al. (2010). "Enhanced psychosocial well-being following participation in a mindfulness-based stress reduction program is associated with increased natural killer cell activity." *The Journal of Alternative and Complementary Medicine*, 16(5), 531–538.

p. 55 the ends of DNA known as the *telomeres*: Epel, E., Daubenmier, J., Moskowitz, J. T., Folkman, S., and Blackburn, E. (2009). "Can meditation slow rate of cellular aging? Cognitive stress, mindfulness, and telomeres." *Annals of the New York Academy of Sciences*, 1172(1), 34–53.

p. 55 deep sleep may also strengthen immunological memories: Westermann, J., Lange, T., Textor, J., and Born, J. (2015). "System consolidation during sleep—A common principle underlying psychological and immunological memory formation." *Trends in Neurosciences*, 38(10), 585–597.

p. 55 Zinc plays an important role in immunity: Rao, G. and Rowland, K. (2011). "Zinc for the common cold—not if, but when." *The Journal of Family Practice*, 60(11), 669.

p. 55 Garlic: Bayan, L., Koulivand, P. H., and Gorji, A. (2014). "Garlic: A review of potential therapeutic effects." *Avicenna Journal of Phytomedicine*, 4(1), 1.

p. 56 long-lasting "memory population": Ataide, Marco A., Komander, Karl, Kastenmüller, Wolfgang, et al. (2020). "BATF3 programs CD8 T cell memory." *Nature Immunology*, 1–11.

4. Mysteries of the Human Cell

p. 59 *G-Protein Pathway Suppressor 2 (GPS2)*: Cardamone, M. D., Tanasa, B., Perissi, V., et al. (2018). "Mitochondrial retrograde signaling in mammals is mediated by the transcriptional cofactor GPS2 via direct mitochondria-to-nucleus translocation." *Molecular Cell*, 69(5), 757–772.

p. 60 Nancy Woolf at the University of California, Los Angeles: Tuszynski, Jack A. *The Emerging Physics of Consciousness*. (Berlin: Springer-Verlag, 2006).

p. 60 maps of about three thousand proteins in yeast cells: Chong, Yolanda, Moffat, Jason, Andrews, Brenda J., et al. (2015), "Yeast Proteome Dynamics from Single Cell Imaging and Automated Analysis." *Cell*, 161, 1413–1424

p. 61 forty-two million protein molecules: Ho, B., Baryshnikova, A., and Brown, G. W. (2017). "Comparative analysis of protein abundance studies to quantify the Saccharomyces cerevisiae proteome." *bioRxiv*, 104919.

p. 61 Atoms measure around one hundred picometers: Atom. In Wikipedia. Retrieved from https://en.wikipedia.org/wiki/Atom.

p. 61 matter particles and force particles: Johnson, George (2016). "Physicists Recover from a Summer's Particle 'Hangover'." *New York Times*.

p. 62 This collective communication is essential to life: Potter, G. D., Byrd, T. A., Mugler, Andrew, and Sun, Bo (2016). "Communication shapes sensory response in multicellular networks." *Proceedings of the National Academy of Sciences*, 113(37), 10334–10339.

p. 62 when our cells need to move: Nordenfelt, P., Elliott, H. L., and Springer, T. A. (2016). "Coordinated integrin activation by actin-dependent force during T-cell migration." *Nature Communications*, 7, 13119; and Marston, D. J., Anderson, K. L., Hanein, D., et al. (2019). "High Rac1 activity is functionally translated into cytosolic structures with unique nanoscale cytoskeletal architecture." *Proceedings of the National Academy of Sciences*, 116(4), 1267–1272.

p. 62 they form contacts with each other: Barone, V., Lang, M., Heisenberg, C. P., et al. (2017). "An effective feedback loop between cell-cell contact duration and morphogen signaling determines cell fate." *Developmental Cell*, 43(2), 198–211.

p. 63 *membrane nanotubes* or *tunneling nanotubes:* Carvalho, R. N. and Gerdes, H. H. (2008). "Cellular Nanotubes: Membrane Channels for Intercellular Communication." In *Medicinal Chemistry and Pharmacological Potential of Fullerenes and Carbon Nanotubes* (Dordrecht, Netherlands: Springer, 2008) 363–372; and Callier, Viviane (2018). "Cells Talk and Help One Another via Tiny Tube Networks." *Quanta Magazine*.

p. 63 new communication system: Murillo, O. D., Thistlethwaite, W., Kitchen, R. R., et al. (2019). "exRNA atlas analysis reveals distinct extracellular RNA cargo types and their carriers present across human biofluids." *Cell*, 177(2), 463–477.

p. 63 new insights into how cells communicate: Ramilowski, J. A., Goldberg, T., Forrest, A. R., et al. (2015). "A draft network of ligand–receptor-mediated multicellular signaling in human." *Nature Communications*, 6.

p. 64 how the transcription factors find their correct places: Yan, Jian, Enge, Martin, Taipale, Minna, Taipale, Jussi, et al. (2013). "Transcription Factor Binding in Human. Cells Occurs in Dense Clusters Formed around Cohesin Anchor Sites." *Cell*, 154 (4), 801.

p. 64 certain cells can remember their earlier encounters: Zarnitsyna, V. I., Huang, J., Zhu, C., et al. (2007). "Memory in receptor–ligand-mediated cell adhesion." *Proceedings of the National Academy of Sciences*, 104(46), 18037–18042.

p. 65 how long a cell's memory lasts: Nasrollahi, S., Walter, C., Pathak, A., et al. (2017). "Past matrix stiffness primes epithelial cells and regulates their future collective migration through a mechanical memory." *Biomaterials*, 146, 146–155.

p. 65 the memory is fully retrievable: Jadhav, U., Cavazza, A., Shivdasani, R. A., et al. (2019). "Extensive recovery of embryonic enhancer and gene memory stored in hypomethylated enhancer DNA." *Molecular Cell*, 74(3), 542–554.

p. 65 New studies in the skin: Peterson, Eric (2018). "Stem Cells Remember Tissues' Past Injuries." *Quanta Magazine*.

p. 66 inflamed skin on mice: Evolution News & Science Today (EN) (2018). "Memory—New Research Reveals Cells Have It, Too." https://evolutionnews.org/2018/11/memory-new-research-reveals-cells-have-it-too/.

p. 66 Shruti Naik: Naik, S., Larsen, S. B., Cowley, C. J., and Fuchs, E. (2018). "Two to tango: dialog between immunity and stem cells in health and disease." *Cell*, 175(4), 908–920.

p. 66 cells somehow remembered the WNT signal: Yoney, A., Etoc, F., Brivanlou, A. H., et al. (2018). "WNT signaling memory is required for ACTIVIN to function as a morphogen in human gastruloids." *eLife*, 7, e38279.

p. 66 neural networks have no monopoly: Blackiston, D. J., Silva, Casey E., and Weiss, M. R. (2008) "Retention of memory through metamorphosis: can a moth remember what it learned as a caterpillar?" *PLoS One* 3(3): e1736.

p. 67 sperm: Alvarez, L., Friedrich, B. M., Gompper, G. and Kaupp, U. B. (2014). "The computational sperm cell." *Trends in Cell Biology*, 24, 198–207.

p. 67 amoebae: Zhu, L., Aono, M., Kim, S.-J. and Hara, M. (2013). "Amoeba-based computing for traveling salesman problem: Long-term correlations between spatially separated individual cells of Physarum polycephalum." *BioSystems*, 112, 1–10.

p. 67 yeast: Caudron, F. and Barral, Y. (2013). "A super-assembly of Whi3 encodes memory of deceptive encounters by single cells during yeast courtship." *Cell*, 155, 1244–57.

p. 67 plants: Grémiaux, A., Yokawa, K., Mancuso, S. and Baluška, F. (2014). Plant anesthesia supports similarities between animals and plants: Claude Bernard's forgotten studies." *Plant Signaling and Behavior*, 9, e27886.

p. 67 bone: Turner, C. H., Robling, A. G., Duncan, R. L., and Burr, D. B. (2002). "Do bone cells behave like a neuronal network?" *Calcified Tissue International*; 70:435–442.

p. 67 heart: Zoghi, M. (2004). "Cardiac memory: Do the heart and the brain remember the same?" *Journal of Interventional Cardiac Electrophysiology*; 11:177–182.

p. 67 bioelectricity in somatic cells: See McCaig, C. D., Rajnicek, A. M., Song, B., and Zhao, M. (2005). "Controlling cell behavior electrically: Current views and future potential." *Physiological Reviews*, 85(3), 943–978; Chakravarthy, S. V. and Ghosh, J. (1997). "On Hebbian-like adaptation in heart muscle: a proposal for 'cardiac memory'." *Biological Cybernetics*, 76(3), 207–215; Inoue, J. (2008). "A simple Hopfield-like cellular network model of plant intelligence." *Progress in Brain Research*, 168, 169–74. See also: Allen, K., Fuchs, E. C., Jaschonek, H., Bannerman, D. M., and Monyer, H. (2011). "Gap Junctions between Interneurons Are Required for Normal Spatial Coding in the Hippocampus and Short-Term Spatial Memory." *Journal of Neuroscience*, 31(17), 6542–6552; Bissiere, S., Zelikowsky, M., Fanselow, M. S., et al. (2011) "Electrical synapses control hippocampal contributions to fear learning and memory." *Science* 331(6013), 87–91; Wu, C. L., Shih, M. F., Chiang, A. S., et al. (2011). "Heterotypic Gap Junctions between Two Neurons in the *Drosophila* Brain Are Critical for Memory." *Current Biology*, 21(10), 848–854; and Tseng, A. and Levin, M. (2013). "Cracking the bioelectric code: Probing endogenous ionic controls of pattern formation." *Communicative & Integrative Biology*, 6(1), 1–8.

p. 67 memory might be distributed throughout the body: Levin, Mike (2013). "Remembrance of Brains Past." *The Node*; retrieved from http://thenode.biologists.com/remembrance-of-brains-past/research/.

p. 67 ways we think sleep is controlled: Ehlen, J. C., Brager, A. J., Joseph S., Takahashi, J., et al. (2017). "Bmal1 function in skeletal muscle regulates sleep." *eLife*, 6.

p. 67 major factor in evolution: Moelling, Karin (2012). "Are viruses our oldest ancestors?" *EMBO Reports*, 13(12), 1033.

p. 68 cell membrane contains receptor and effector proteins: Lipton, B. H. (2001). "Insight into cellular consciousness," *Bridges*, 12(1), 5.

p. 68 organic *computer chip* or *microchip*: Lipton, B. H. *The Biology of Belief: Unleashing the Power of Consciousness, Matter & Miracles.* (Santa Rosa, Calif.: Mountain Of Love/Elite Books, 2005).

p. 68 *nanomicroscope*: Contera, S. *Nano Comes to Life: How Nanotechnology Is Transforming Medicine and the Future of Biology.* (Princeton, N.J.: Princeton University Press, 2019).

p. 69 biological cell membrane into a digital readout computer chip: Cornell, B. A., Braach-Maksvytis, V., Pace R., et al. (1997). "A biosensor that uses ion-channel switches." *Nature*, 387, 580–583.

p. 69 "imitating highly sophisticated biological neural networks": Nili, Hussein, Walia, Sumeet, Sriram, Sharath, et al. (2015). "Donor-Induced Performance Tuning of Amorphous SrTiO3 Memristive Nanodevices: Multistate Resistive Switching and Mechanical Tunability." *Advanced Functional Materials*, 25(21), 3172–3182.

p. 70 one million pictures or one hundred hours of 3D movies: Yang Fu-liang in Savage, Sam (2010). "Scientists Create World's Smallest Microchip." RedOrbit.com, https://www.redorbit.com/news/technology/1966110/scientists_create_worlds_smallest_microchip/.

p. 70 Michigan Micro Mote: June, Catharine (2015). "Michigan Micro Mote (M3) makes history as the world's smallest computer." University of Michigan. https://ece.engin.umich.edu/stories/michigan-micro-mote-m3-makes-history-as-the-worlds-smallest-computer.

p. 70 *smart dust*: Templeton, G. (2014). "Smart dust: a complete computer that's smaller than a grain of sand." *ExtremeTech*. Retrieved from: http://www.extremetech.com/extreme/155771-smart-dust-a-complete-computer-thats-smaller-than-a-grain-of-sand; accessed: Jan. 30, 2014.

p. 70 smallest memory device yet: Hus, S. M., Ge, R., Akinwande, D., et al. (2020). "Observation of single-defect memristor in an MoS2 atomic sheet." *Nature Nanotechnology*, 1–5.

p. 70 engineering circuits into bacterial cells: Chen, A. Y., Zhong, C., Lu, T. K. (2015). "Engineering living functional materials." *ACS Synthetic Biology*, 4(1), 8–11.

p. 70 encode all of Shakespeare's sonnets into DNA: Goldman, N., Bertone, P., Chen, S., Dessimoz, C., LeProust, E. M., Sipos, B., and Birney, E. (2013). "Towards practical, high-capacity, low-maintenance information storage in synthesized DNA." *Nature*, 494(7435), 77–80.

p. 71 within the genomes of populations of living cells: Shipman, Seth L., Nivala, Jeff, Macklis, Jeffrey D., and Church, George M. (2017). "CRISPR–Cas encoding of a digital movie into the genomes of a population of living bacteria." *Nature*, nature23017.

p. 71 George Church, a geneticist at Harvard University: Church, G. M., Gao, Y., and Kosuri, S. (2012). "Next-generation digital information storage in DNA." *Science*, 1226355.

5. The Intelligence of Single-Celled Organisms

p. 74 to differentiate between plants, organisms, and animals: The difference between organism and animal: In *Wikipedia*: http://wikidiff.com/organism/animal.

p. 74 what constitutes intelligence: Humphreys, Lloyd G. (1979). "The construct of general intelligence." *Intelligence,* 3(2), 105–120.

p. 74 goal-directed behavior, and memory: Guan, Q., Haroon, S., Gasch, A. P., et al. (2012). "Cellular memory of acquired stress resistance in Saccharomyces cerevisiae." *Genetics,* 192(2), 495–505.

p. 75 They sense and explore their surroundings: Rennie, John and Reading-Ikkanda, Lucy (2017) "Seeing the Beautiful Intelligence of Microbes." *Quanta Magazine.*

p. 75 all plants and animals combined: Bacteria: In *Wikipedia.* https://en.wikipedia.org /wiki/Bacteria.

p. 75 researchers at McGill University have discovered bacterial organelles: Ladouceur, A. M., Parmar, B., Weber, S. C., et al. (2020). "Clusters of bacterial RNA polymerase are biomolecular condensates that assemble through liquid-liquid phase separation." *bioRxiv.*

p. 76 enzyme called *Cas9*: Heler, R., Samai, P., Marraffini, L. A., et al. (2015). "Cas9 specifies functional viral targets during CRISPR-Cas adaptation." *Nature,* 519(7542), 199.

p. 76 "remembering" prior infections: Nuñez, J. K., Lee, A. S., Engelman, A., and Doudna, J. A. (2015). "Integrase-mediated spacer acquisition during CRISPR–Cas adaptive immunity." *Nature*, 519(7542), 193.

p. 76 Other researchers at Berkeley: Nuñez, J. K., Lee, A. S., Engelman, A., and Doudna, J. A. (2015). "Integrase-mediated spacer acquisition during CRISPR–Cas adaptive immunity." *Nature*, 519(7542), 193.

p. 76 encoded in their organisms: Wolf, D. M., Fontaine-Bodin, L., Arkin, A. P., et al. (2008). "Memory in microbes: quantifying history-dependent behavior in a bacterium." *PLoS One*, 3(2), e1700.

p. 77 Pak Chung Wong at the Pacific Northwest National Laboratory: Vaidyanathan, G. (2017). "Science and Culture: Could a bacterium successfully shepherd a message through the apocalypse?" *Proceedings of the National Academy of Sciences*, 114(9), 2094–2095.

p. 77 *quorum sensing*: Kim, M. K., Ingremeau, F., Zhao, A., Bassler, B. L., and Stone, H. A. (2016). "Local and global consequences of flow on bacterial quorum sensing." *Nature Microbiology,* 1(1), 1–5.

p. 77 operate in an interspecies manner: Dubey, G. P. and Ben-Yehuda, S. (2011). "Intercellular nanotubes mediate bacterial communication." *Cell,* 144(4), 590–600.

p. 78 bacteria use *ion channels* to communicate: Prindle, A., Liu, J., Süel, G. M., et al. (2015). "Ion channels enable electrical communication in bacterial communities." *Nature,* 527(7576), 59–63.

p. 78 Scientists from MIT: Gandhi, S. R., Yurtsev, E. A., Korolev, K. S., and Gore, J. (2016). "Range expansions transition from pulled to pushed waves as growth becomes more cooperative in an experimental microbial population." *Proceedings of the National Academy of Sciences,* 113(25), 6922–6927.

p. 78 bacteria are equipped with sensory systems: Koshland Jr., D. E. (1980). "Bacterial chemotaxis in relation to neurobiology." *Annual Review of Neuroscience*, 3(1), 43–75; and Lyon, P. (2015). "The cognitive cell: bacterial behavior reconsidered." *Frontiers in Microbiology*, 6. and https://www.flinders.edu.au/biofilm-research-innovation -consortium.

p. 78 Roberto Kolter: Chimileski, Scott and Kolter, Roberto (2018). "Microbial Life: A Universe at the Edge of Sight." https://hmnh.harvard.edu/microbial-life-universe-edge-sight.

p. 78 *microbial brain*: Baluška, F. and Mancuso, S. (2009). "Deep evolutionary origins of neurobiology: Turning the essence of 'neural' upside-down." *Communicative & Integrative Biology*, 2(1), 60–65.

p. 78 ancient remnant of symbiotic bacteria: Landau, Elizabeth (2020). "Mitochondria May Hold Keys to Anxiety and Mental Health." *Quanta Magazine*.

p. 79 development and function of the immune system: Sampson, T. R. and Mazmanian, S. K. (2015). "Control of brain development, function, and behavior by the microbiome." *Cell Host & Microbe*, 17(5), 565–576.

p. 79 brain metabolites: Clarke, G., O'Mahony, S. M., Dinan, T. G., and Cryan, J. F. (2014). "Priming for health: gut microbiota acquired in early life regulates physiology, brain and behaviour." *Acta Paediatrica*, 103(8), 812–819; and Janik, R., Thomason, L. A., Stanisz, A. M., Stanisz, G. J., et al. (2016). "Magnetic resonance spectroscopy reveals oral Lactobacillus promotion of increases in brain GABA, N acetyl aspartate and glutamate." *Neuroimage*, 125, 988–995.

p. 79 behavior: Sampson, T. R. and Mazmanian, S. K. (2015). "Control of brain development, function, and behaviour by the microbiome." *Cell Host & Microbe*, 17(5), 565–576.

p. 79 neurogenesis: Ogbonnaya, E. S., Clarke, G., and O'Leary, O. F. (2015). "Adult hippocampal neurogenesis is regulated by the microbiome." *Biological Psychiatry*, 78(4), e7–e9.

p. 79 Gut microbes contain 3.3 million genes: Qin, J., Li, R., Mende, D. R., et al. (2010). "A human gut microbial gene catalogue established by metagenomic sequencing." *Nature*, 464(7285), 59–65.

p. 79 (SCFAs) that are essential for host health: Russell, W. R., Hoyles, L., Flint, H. J., and Dumas, M. E. (2013). "Colonic bacterial metabolites and human health." *Current Opinion in Microbiology*, 16(3), 246–254; and Tan, J., McKenzie, C., Potamitis, M., Macia, L., et al. (2014). "The role of short-chain fatty acids in health and disease." *Advances in Immunology*, 121(91), e119.

p. 79 vagus nerve: Grenham, S., Clarke, G, Dinan, T. G., et al. (2011). "Brain–gut–microbe communication in health and disease." *Frontiers in Physiology*, 2; and Montiel-Castro, A. J., González-Cervantes, R. M., and Pacheco-López, G. (2013). "The microbiota-gut-brain axis: neurobehavioral correlates, health and sociality." *Frontiers in Integrative Neuroscience*, 7.

p. 80 neuronal reward pathway: Han, W., Tellez, L. A., Kaelberer, M. M., et al. (2018). "A neural circuit for gut-induced reward." *Cell*, 175(3), 665–678.

p. 80 right and left branches of the vagus nerve: Han, W., Tellez, L. A., Kaelberer, M. M., et al. (2018). "A neural circuit for gut-induced reward." *Cell*, 175(3), 665–678.

p. 80 "The Second Brain": Gershon, Michael D. (2020). "How smart is the gut?" *Acta Physiologica*. 228, e13296; Gershon, Michael D. *The Second Brain*. (New York: HarperCollins, 1998); Rao, M. and Gershon, M. D. (2017). "The dynamic cycle of life in the enteric nervous system." *Nature Reviews Gastroenterology & Hepatology*, 14(8), 453–454; and Mungovan, K. and Ratcliffe, E. M. (2016). "Influence of the Microbiota on the Development and Function of the 'Second Brain'—The Enteric Nervous System." In *The Gut-Brain Axis*. (Cambridge, Mass.: Academic Press, 2016), 403–421.

p. 80 the ENS represents a "smart" system: Schemann, M., Frieling, T. and Enck, P. (2020). "To learn, to remember, to forget—How smart is the gut?" *Acta Physiologica*, 228(1), e13296.

p. 80 depression: Dinan, T. G. and Cryan, J. F. (2013). "Melancholic microbes: a link between gut microbiota and depression." *Neurogastroenterology and Motility*, 25(9), 713–719.

p. 80 irritable bowel syndrome: Vandvik, P. O., Wilhelmsen, I., and Farup, P. G. (2004). "Comorbidity of irritable bowel syndrome in general practice: A striking feature with clinical implications." *Alimentary Pharmacology & Therapeutics*, 20(10), 1195–1203; and Dinan, T. G. and Cryan, J. F. (2013). "Melancholic microbes: a link between gut microbiota and depression?" *Neurogastroenterology and Motility*, 25(9), 713–719.

p. 81 Gastrointestinal problems: "The rise of 'psychobiotics'? 'Poop pills' and probiotics could be game changers for mental illness." Sharon Kirkey. *National Post*, October 8, 2019.

p. 81 psychotropic drugs: European College of Neuropsychopharmacology. "Scientists find psychiatric drugs affect gut contents." *ScienceDaily*, September 9, 2019.

p. 81 diet rich in salt: Faraco, G., Brea, D., Sugiyama, Y., et al. (2018). "Dietary salt promotes neurovascular and cognitive dysfunction through a gut-initiated TH17 response." *Nature Neuroscience*, 21(2), 240–249.

p. 82 important contributions to our health: Hu, S., Dong, T. S. and Chang, E. B. (2011). "The microbe-derived short chain fatty acid butyrate targets miRNA-dependent p21 gene expression in human colon cancer." *PloS One*, *6*(1), e16221; and Shenderov, B. A. (2012). "Gut indigenous microbiota and epigenetics." *Microbial Ecology in Health and Disease*, 23(1), 17195.

p. 82 gut bacteria on the aging process: Hartsough, L. A., Kotlajich, M. V., Tabor, J. J., et al. (2020). "Optogenetic control of gut bacterial metabolism." *eLife*.

p. 82 initial colonization of their gut: Younge, N., McCann, J. R. and Seed, P. C., et al. (2019). "Fetal exposure to the maternal microbiota in humans and mice." *JCI Insight*.

p. 82 mother's birth canal: Koenig, J. E., Spor, A., Ley, R. E., et al. (2011). "Succession of microbial consortia in the developing infant gut microbiome." *Proceedings of the National Academy of Sciences*, 108(supplement 1), 4578–4585.

p. 82 Infants delivered by caesarean section: Faa, G., Gerosa, C., Fanos, V., et al. (2013). "Factors influencing the development of a personal tailored microbiota in the neonate, with particular emphasis on antibiotic therapy." *The Journal of Maternal-Fetal & Neonatal Medicine*, 26(sup2), 35–43.

p. 83 BPA exposure just before or after birth: Reddivari, Lavanya, Veeramachaneni, D. N., Rao, Vanamala, Jairam, K. P., et al. (2017). "Perinatal Bisphenol A Exposure Induces Chronic Inflammation in Rabbit Offspring via Modulation of Gut Bacteria and Their Metabolites." *mSystems*, 2(5).

p. 83 direct cell-to-cell communication: Prindle, A., Liu, J., Süel, G. M., et al. (2015). "Ion channels enable electrical communication in bacterial communities." *Nature*, 527(7576), 59–63.

p. 83 associated with changes in thinking, feeling, and behaving: Allman, J. M. *Evolving Brains*. (New York: Scientific American Library, 1999); Damasio, A. R. (1999). "The feeling of what happens: Body and emotion in the making of consciousness." *New York Times Book Review*, 104, 8–8; and Greenspan, R. J. *An Introduction to Nervous Systems*. (Cold Spring Harbor, N.Y.: Cold Spring Harbor Laboratory Press, 2007).

p. 83 Slime molds . . . a billion years ago: Sandra Baldauf in Zimmer, Carl (2011). "Can Answers to Evolution Be Found in Slime?" *New York Times*.

p. 84 *The Social Amoebae:* Bonner, John. *The Social Amoebae: The Biology of the Cellular Slime Molds*. (Princeton, N.J.: Princeton University Press, 2009).

p. 84 digesting the engulfed material: McDonnell Genome Institute, http://genome.wustl.edu/genomes/detail/physarum-polycephalum/.

p. 84 basic behavioral patterns: Whiting, J. G., Jones, J., Bull, L., Levin, M., and Adamatzky, A. (2016). "Towards a Physarum learning chip." *Scientific Reports,* 6.

p. 84 spatial memory system in a nonneuronal organism: Reid, C. R., MacDonald, H., Mann, R. P., Marshall, J. A., Latty, T., and Garnier, S. (2016). "Decision-making without a brain: How an amoeboid organism solves the two-armed bandit." *Journal of the Royal Society Interface,* 13(119), 20160030; and Chris Reid's web page: http://sydney.edu.au/science/biology/socialinsects/profiles/chris-reid.shtml; and Smith-Ferguson, J., Reid, C. R., Latty, T., and Beekman, M. (2017). "Hänsel, Gretel and the slime mould–how an external spatial memory aids navigation in complex environments." *Journal of Physics D: Applied Physics*, 50(41), 414003.

p. 84 This simple organism has the ability: Nakagaki, T., Yamada, H., and Hara, M. (2004). "Smart network solutions in an amoeboid organism." *Biophysical Chemistry,* 107(1), 1–5; and Shirakawa, T. and Gunji, Y. P. (2007). "Emergence of morphological order in the network formation of Physarum polycephalum." *Biophysical Chemistry,* 128(2), 253–260.

p. 84 solve mazes: Adamatzky, A. and Jones, J. (2010). "Road planning with slime mould: if Physarum built motorways it would route M6/M74 through Newcastle." *International Journal of Bifurcation and Chaos,* 20(10), 3065–3084; and Nakagaki, T., Yamada, H., and Tóth, Á. (2000). "Intelligence: Maze-solving by an amoeboid organism." *Nature,* 407(6803), 470.

p. 84 approximate human transport networks: Adamatzky, A. I. (2014). "Route 20, autobahn 7, and slime mold: approximating the longest roads in USA and Germany with slime mold on 3-D terrains." *IEEE Transactions on Cybernetics,* 44(1), 126–136.

p. 85 long-term memory: Zhu, L., Aono, M., Kim, S. J., and Hara, M. (2013). "Amoeba-based computing for traveling salesman problem: Long-term correlations between spatially separated individual cells of Physarum polycephalum." *BioSystems*, 112(1), 1–10.

p. 85 plasmodium tube network: Nakagaki, T., Yamada, H., and Hara, M. (2004). "Smart network solutions in an amoeboid organism." *Biophysical Chemistry,* 107(1), 1–5.

p. 85 emergent intelligence: Tsuda, S., Aono, M., and Gunji, Y. P. (2004). "Robust and emergent Physarum logical-computing." *BioSystems,* 73(1), 45–55.

p. 85 cellular origins of primitive intelligence: Saigusa, T., Tero, A., Nakagaki, T., and Kuramoto, Y. (2008). "Amoebae anticipate periodic events." *Physical Review Letters,* 100(1), 018101.

p. 85 "this paper would add a cellular memory": James Shapiro in Ball, P. (2008). "Cellular memory hints at the origins of intelligence." *Nature* 451, 385.

p. 85 When food is in short supply during pregnancy: Blaser, Martin J. (2014). "The Way You're Born Can Mess with the Microbes You Need to Survive." www.wired.

p. 86 mice treated with antibiotics: Brown, D. G., Soto, R., Fujinami, R. S., et al. (2019). "The microbiota protects from viral-induced neurologic damage through microglia-intrinsic TLR signaling." *eLife,* 8, e47117.

p. 86 having a healthy and diverse microbiota is crucial: Brown, D. G., Soto, R. and Fujinami, R. S. (2019). "The microbiota protects from viral-induced neurologic damage through microglia-intrinsic TLR signaling." *eLife,* 8, e47117.

p. 86 Two studies: Allen, J. M., Mailing, L. J, Woods, J. A., et al. (2018). "Exercise alters gut microbiota composition and function in lean and obese humans." *Medicine and Science in Sports and Exercise*, 50(4), 747–757.

p. 86 people who ate avocados every day: Thompson, S. V., Bailey, M. A. and Holscher, H. D. (2020). "Avocado Consumption Alters Gastrointestinal Bacteria Abundance and Microbial Metabolite Concentrations among Adults with Overweight or Obesity: A Randomized Controlled Trial." *The Journal of Nutrition*.

p. 87 optimal vitamin D concentration: Thomas, R. L., Jiang, L., Orwoll, E. S., et al. (2020). "Vitamin D metabolites and the gut microbiome in older men." *Nature Communications*, 11(1), 1–10.

p. 87 Studies of cognition: Bräuer, J., Hanus, D., and Uomini, N. (2020). "Old and New Approaches to Animal Cognition: There Is Not 'One Cognition.'" *Journal of Intelligence*, 8(3), 28.

p. 88 microbiome has multiple critical effects: Sommer, F. and Bäckhed, F. (2013). "The gut microbiota-masters of host development and physiology." *Nature Reviews Microbiology*, 11(4), 227.

6. Regeneration, Hibernation, and Metamorphosis

p. 91 small slices of the adult worm: Umesono, Y., Tasaki, J., Nishimura, K., Inoue, T., and Agata, K. (2011). "Regeneration in an evolutionarily primitive brain—the planarian Dugesia japonica model." *European Journal of Neuroscience*, 34, 863–869.

p. 91 species and the availability of food: Baldscientist (2011). "Playing With Worms." Retrieved from https://baldscientist.wordpress.com/2011/09/23/playing-with-wormies/.

p. 91 after surgical removal of that part: Gentile, L., Cebria, F. and Bartscherer, K. (2011) "The planarian flatworm: an in vivo model for stem cell biology and nervous system regeneration." *Disease Models & Mechanisms*, 4(1), 12–9; and Lobo, D., Beane, W. S., and Levin, M. (2012). "Modeling planarian regeneration: A primer for reverse-engineering the worm," *PLoS Computational Biology*, 8(4), e1002481.

p. 92 popular organism for the study of memory: Sarnat, H. B. and Netsky, M. G. (1985). "The brain of the planarian as the ancestor of the human brain." *Canadian Journal of Neurological Sciences*, 12(4), 296–302.

p. 92 distributed throughout the animal's body: McConnell, J. V., Jacobson, A. L., and Kimble, D. P. (1959). "The effects of regeneration upon retention of a conditioned response in the planarian." *Journal of Comparative and Physiological Psychology*, 52, 1–5.

p. 92 group led by Tal Shomrat and Mike Levin: Shomrat, T. and Levin, M. (2013). "An automated training paradigm reveals long-term memory in planarians and its persistence through head regeneration." *The Journal of Experimental Biology*, 216(20), 3799–3810.

p. 92 Michael Levin represents a unique hire: Robert Sternberg in Tufts University News (2008). "Biologist Michael Levin Joins Tufts University." News release, retrieved from http://now.tufts.edu/news-releases/biologist-michael-levin-joins-tufts-university.

p. 93 2013 paradigm-shifting paper: Shomrat, T. and Levin, M. (2013). "An automated training paradigm reveals long-term memory in planarians and its persistence through head regeneration." *The Journal of Experimental Biology*, 216(20), 3799–3810.

p. 93 suffices to restore full memory: Inoue, T., Kumamoto, H., Okamoto, K., Umesono, Y., Sakai, M., Sánchez Alvarado, A., and Agata, K. (2004). "Morphological and functional recovery of the planarian photosensing system during head regeneration." *Zoological Science*, 21, 275–283

p. 94 regenerative and developmental processes observed in the planaria: Levin, M. (2013). "Reprogramming cells and tissue patterning via bioelectrical pathways: Molecular mechanisms and biomedical opportunities." *Wiley Interdisciplinary Reviews: Systems*

Biology and Medicine, 5:657–676; Thurler, K. (2013). "Flatworms lose their heads but not their memories." Retrieved from: https://now.tufts.edu/news-releases/flatworms-lose-their-heads-not-their-memories; Mustard, J. and Levin, M. (2014). "Bioelectrical mechanisms for programming growth and form: Taming physiological networks for soft body robotics." *Soft Robotics*, 1(3), 169–191; Blackiston, D. J., Silva, Casey E., and Weiss, M. R. (2008) "Retention of Memory through Metamorphosis: Can a Moth Remember What It Learned As a Caterpillar?" *PLoS One* 3(3), e1736; and Blackiston, D. J., Shomrat, T., and Levin, M. (2015). "The stability of memories during brain remodeling: A perspective." *Communicative & Integrative Biology*, 8(5), e1073424.

p. 94 depending on the species: Humphries, M. M., Thomas, D.W., and Kramer, D.L. (2003). "The role of energy availability in mammalian hibernation: A cost-benefit approach." *Physiological and Biochemical Zoology*, 76(2), 165–179

p. 94 In an Austrian study ground squirrels were trained: Millesi, E., Prossinger, H., Dittami, J. P., and Fieder, M. (2001). "Hibernation effects on memory in European ground squirrels (Spermophilus citellus)." *Journal of Biological Rhythms*, 16, 264–271;

p. 95 selectively initiating *autophagy*: Hadj-Moussa, H. and Storey, K. B. (2018). "Micromanaging freeze tolerance: the biogenesis and regulation of neuroprotective microRNAs in frozen brains." *Cellular and Molecular Life Sciences*, 75(19), 3635–3647.

p. 95 Shrews: Lázaro, J., Dechmann, D. K., LaPoint, S., Wikelski, M., and Hertel, M. (2017). "Profound reversible seasonal changes of individual skull size in a mammal." *Current Biology*, 27(20), R1106–R1107.

p. 95 bats: Ruczynski, I. and Siemers, B. M. (2011). "Hibernation does not affect memory retention in bats." *Biology Letters*, 7(1), 153–155.

p. 96 hibernation in Alpine marmots: Clemens, L. E., Heldmaier, G., and Exner, C. (2009). "Keep cool: Memory is retained during hibernation in Alpine marmots." *Physiology & Behavior*, 98(1), 78–84.

p. 96 wood frogs: Sullivan, K. J. and Storey, K. B. (2012). "Environmental stress responsive expression of the gene li16 in *Rana sylvatica*, the freeze tolerant wood frog." *Cryobiology*, 64, 192–200.

p. 96 *Holometabolous insects*: Sheiman, I. M. and Tiras, K. L. (1996). "Memory and morphogenesis in planaria and beetle," in C. I. Abramson, Z. P. Shuranova, and Y. M. Burmistrov (eds.) *Russian Contributions to Invertebrate Behavior* (Westport, Conn.: Praeger, 1996); and Blackiston, D. J., Silva, Casey E., and Weiss, M. R. (2008). "Retention of memory through metamorphosis: can a moth remember what it learned as a caterpillar?," *PLoS One* 3(3), e1736

p. 97 The adult moth's brain: Synnøve Ressem (2010). "Inside a moth's brain." *Gemini Magazine*.

p. 97 team of researchers at Tufts University: Blackiston, D. J., Casey, E. S., and Weiss, M. R. (2008). "Retention of memory through metamorphosis: Can a moth remember what it learned as a caterpillar?" *PLoS One*, 3(3), e1736.

p. 97 Two possible mechanisms: Blackiston, D. J., Silva, Casey E., Weiss, M. R. (2008). "Retention of Memory through Metamorphosis: Can a Moth Remember What It Learned As a Caterpillar?" *PLoS One*, 3(3), e1736.

p. 97 Yukihisa Matsumoto at Hokkaido University: Matsumoto, Y. and Mizunami, M. (2002) "Lifetime olfactory memory in the cricket *Gryllus bimaculatus*." *Journal of Comparative Physiology A-Neuroethology Sensory Neural and Behavioral Physiology* 188: 295–299.

p. 98 "embryonic" learning in amphibians: Hepper, P. G. and Waldman, B. (1992). "Embryonic olfactory learning in frogs." *Quarterly Journal of Experimental Psychology*, Section B, 44(3–4), 179–197.

p. 98 Working with wood frog larvae: Mathis, A., Ferrari, M. C., Windel, N., Messier, F., and Chivers, D. P. (2008). "Learning by embryos and the ghost of predation future." *Proceedings Biological Sciences*, 275, 2603–2607.

p. 98 beetles: Rietdorf, K. and Steidle, J.L.M. (2002) "Was Hopkins right? Influence of larval and early adult experience on the olfactory response in the granary weevil *Sitophilus granarius* (Coleoptera, Curculionidae)". *Physiological Entomology*, 27, 223–227.

p. 98 fruit flies: Trewavas, A. (2002). "Plant intelligence: Mindless mastery." *Nature*, 415(6874), 841–841.

p. 98 ants: Trewavas, A. (a2005). "Green plants as intelligent organisms." *Trends in Plant Science*, 10(9), 413–419.

p. 98 parasitic wasps: Trewavas, A. (b2005). "Plant intelligence." *Naturwissenschaften*, 92(9), 401–413.

p. 98 Neuroscientist May-Britt Moser: Synnøve Ressem (2010). "Inside a moth's brain." *Gemini Magazine.*

7. The Wisdom of the Body

p. 102 "self-remembering systems": Dudas, M., Wysocki, A., Gelpi, B., and Tuan, T. L. (2008). "Memory encoded throughout our bodies: molecular and cellular basis of tissue regeneration." *Pediatric Research*, 63(5), 502–51.

p. 102 "down at the cellular level": van der Kolk, B. A. (1994). "The body keeps the score: Memory and the evolving psychobiology of posttraumatic stress." *Harvard Review of Psychiatry*, 1(5), 253–265.

p. 102 *The Inner Game of Tennis*: Gallwey, W. Timothy. *The Inner Game of Tennis: The Classic Guide to the Mental Side of Peak.* (New York: Random House, 1997).

p. 103 central nervous system: Rutherford, O. M. and Jones, D. A. (1986). "The role of learning and coordination in strength training." *European Journal of Applied Physiology and Occupational Physiology*, 55, 100–105.

p. 103 muscle memory: Staron, R. S., Leonardi, M. J., Karapondo, D. L., Malicky, E. S., Falkel, J. E., Hagerman, F. C., and Hikida, R. S. (1991). "Strength and skeletal muscle adaptations in heavy-resistance-trained women after detraining and retraining." *Journal of Applied Physiology*, 70, 631–640; Taaffe, D. R. and Marcus, R. (1997). "Dynamic muscle strength alterations to detraining and retraining in elderly men." *Clinical Physiology*, 17, 311–324.

p. 103 permanent once they are formed: Egner, I. M., Bruusgaard, J. C., Eftestøl, E., and Gundersen, K. (2013). "A cellular memory mechanism aids overload hypertrophy in muscle long after an episodic exposure to anabolic steroids." *The Journal of Physiology*, 591(24), 6221–6230.

p. 103 producing more mitochondria: Lee, Hojun, Kim, Boa, Kawata, Keisuke, et al. (2015). "Cellular Mechanism of Muscle Memory: Effects on Mitochondrial Remodeling and Muscle Hypertrophy." *Medicine and Science in Sports and Exercise*, 47(5S), 101–102; www.sciencedaily.com/releases/2017/08/170803145629.htm.

p. 104 sleep-wake rhythm in humans: Meyer, K., Köster, T., Staiger, D., et al. (2017). "Adaptation of iCLIP to plants determines the binding landscape of the clock-regulated RNA-binding protein At GRP7." *Genome Biology*, 18(1), 204.

p. 104 circadian clock protein in the muscle—BMAL1: UT Southwestern Medical Center (2017). "Muscle, not brain, may hold answers to some sleep disorders." *ScienceDaily.*

p. 104 molecular mechanisms controlling the circadian rhythm: Nobel Prize in Medicine, 2017, https://www.nobelprize.org/nobel_prizes/medicine/laureates/2017/.

p. 105 sex difference in circadian alignment: Duffy, J. F., Cain, S. W., Chang, A. M., Phillips, A. J., Münch, M. Y., Gronfier, C., and Czeisler, C. A. (2011). "Sex difference in the near-24-hour intrinsic period of the human circadian timing system." *Proceedings of the National Academy of Sciences*, 108 (supplement 3), 15602–15608.

p. 105 cancer prevention: El-Athman, R., Genov, N. N., Relógio, A., et al. (2017). "The Ink4a/Arf locus operates as a regulator of the circadian clock modulating RAS activity." *PLoS Biology*, 15(12), e2002940.

p. 105 influence of microenvironmental conditions: Dudas, M., Wysocki, A., Gelpi, B., and Tuan, T. L. (2008). "Memory encoded throughout our bodies: molecular and cellular basis of tissue regeneration." *Pediatric Research*, 63(5), 502–512.

p. 106 researchers from Rockefeller University: Naik, S., Larsen, S. B., Fuchs, E., et al. (2017). "Inflammatory memory sensitizes skin epithelial stem cells to tissue damage." *Nature*, 550(7677), 475–480.

p. 107 new study from King's College: Denk, F., Crow, M., Didangelos, A., Lopes, D. M. and McMahon, S. B. (2016). "Persistent alterations in microglial enhancers in a model of chronic pain." *Cell Reports*, 15(8), 1771–1781.

p. 107 original function of a brain area: Kikkert, S., Kolasinski, J., Makin, T. R., et al. (2016). "Revealing the neural fingerprints of a missing hand." *eLife*, 5, e15292.

p. 108 William James: James, William (1884). "What Is an Emotion?" *Mind*, os-IX(34), 188–205; and James, William. "What is an Emotion" in Richardson, R. D. (ed.). *The Heart of William James*. (Cambridge, Mass., 2012).

p. 108 Marc Wittman: Wittmann, Marc. *Felt Time: The Psychology of How We Perceive Time*. (Cambridge, Mass.: MIT Press, 2016); and Wittmann, Marc (2018). *Altered States of Consciousness: Experiences Out of Time and Self.* (Cambridge, Mass.: MIT Press, 2016)

p. 108 Wilhelm Reich: Oofana, Ben (April 15, 2018). "Dissolving the Layers of Emotional Body Armor." https://benoofana.com/dissolving-the-layers-of-emotional-body-armor.

p. 109 Health professionals such as osteopaths: Upledger, John E. *Your Inner Physician and You: Craniosacral Therapy and Somatoemotional Release*. (Berkeley, Calif.: Atlantic Books, 1997).

p. 110 *post-traumatic stress disorder* (PTSD): American Psychiatric Association. (2013). *Diagnostic and statistical manual of mental disorders (DSM-5)*. (Washington: American Psychiatric Publishing, 2013).

p. 110 "*The war sits in all my bones*": Appelfeld, A. (2005). *Geschichte eines Lebens*. (Berlin: Rowohlt, 2005).

p. 110 Traumatized persons react to reminders: Fuchs, T. (2012). "Body memory and the unconscious. In *Founding Psychoanalysis Phenomenologically* (Dordrecht, Netherlands: Springer, 2012), 69–82.

p. 110 acute stress response in bony vertebrates is not possible without *osteocalcin*: Obri, A., Khrimian, L., Karsenty, G., and Oury, F. (2018). "Osteocalcin in the brain: from embryonic development to age-related decline in cognition." *Nature Reviews Endocrinology*, 14(3), 174.

p. 111 "I think what that means . . .": Herman, James (September 12, 2019). In "Fight or Flight May Be in Our Bones" by Diana Kwon. *Scientific American*.

p. 111 fat-to-brain signal: de Kloet, A. D. and Herman, J. P. (2018). "Fat-brain connections: Adipocyte glucocorticoid control of stress and metabolism." *Frontiers in Neuroendocrinology*, 48, 50–57.

p. 111 no organ, including the brain, is an island: Obri, A., Khrimian, L., Karsenty, G., and Oury, F. (2018). "Osteocalcin in the brain: from embryonic development to age-related decline in cognition." *Nature Reviews Endocrinology*, 14(3), 174.

p. 111 ability to generate new nuclei in muscles: Egner, I. M., Bruusgaard, J. C., Eftestøl, E., and Gundersen, K. (2013). "A cellular memory mechanism aids overload hypertrophy in muscle long after an episodic exposure to anabolic steroids." *The Journal of Physiology*, 591(24), 6221–6230.

p. 111 people's responses to extreme experiences: American Psychiatric Association. *Diagnostic and statistical manual of mental disorders* (4th edition), (Washington: American Psychiatric Press, 2000); and van der Kolk, B. A. *Psychological Trauma*. (Washington: American Psychiatric Press, 1987).

p. 112 Body-Centered Psychotherapy: Fogel, Alan (2010). "PTSD is a chronic impairment of the body sense: Why we need embodied approaches to treat trauma," in *Body Sense*, online blog from *Psychology Today*. https://www.psychologytoday.com/blog/body-sense201006/ptsd-is-chronic-impairment-the-body-sense-why-we-need-embodied-approaches.

p. 112 "they were 'buried' in our parents' lives": Earnshaw, Averil. *Time Will Tell*. (Sidney, Australia: A and K Enterprises, 1995).

8. Heart Transplants: Personality Transplants?

p. 118 generates a tiny electrical current: Merck Manual. Retrieved from http://www.merckmanuals.com/en-ca/home/heart-and-blood-vessel-disorders/biology-of-the-heart-and-blood-vessels/biology-of-the-heart.

p. 118 pluripotent cardiac stem cells: Bu, L., Jiang, X., Chien, K. R., et al. (2009). "Human ISL1 heart progenitors generate diverse multipotent cardiovascular cell lineages." *Nature*, 460(7251), 113.

p. 118 constantly interact with each other: Tirziu, D., Giordano, F. J., and Simons, M. (2010). "Cell communications in the heart." *Circulation*, 122(9), 928–937.

p. 118 intrinsic nervous system of the heart: Armour, J., "Anatomy and function of the intrathoracic neurons regulating the mammalian heart." *Reflex Control of the Circulation*, I. H. Zucker and J. P. Gilmore, eds. (Boca Raton, Fla.: CRC Press, 1991). 1–37; and Armour, J. A. (2008). "Potential clinical relevance of the 'little brain' on the mammalian heart." *Experimental Physiology*, 93(2), 165–176.

p. 119 contract as a single unit: Chakravarthy, S. V. and Ghosh, J. (1997). "On Hebbian-like adaptation in heart muscle: a proposal for 'cardiac memory'." *Biological Cybernetics*, 76(3), 207–215.

p. 119 neurotransmitter acetylcholine: Merck Manual. Retrieved from http://www.merckmanuals.com/en-ca/home/heart-and-blood-vessel-disorders/biology-of-the-heart-and-blood-vessels/biology-of-the-heart.

p. 119 study from the University of Buenos Aires: Rosenbaum, M. B., Sicouri, S. J., Davidenko, J. M., and Elizari, M. V. (1985). "Heart rate and electrotonic modulation of the T wave: a singular relationship." *Cardiac Electrophysiology and Arrhythmias*. (New York: Grune and Stratton, 1985).

p. 119 properties of memory and adaptation: Chakravarthy, S. V. and Ghosh, J. (1997). "On Hebbian-like adaptation in heart muscle: a proposal for 'cardiac memory'." *Biological Cybernetics*, 76(3), 207–215.

p. 120 heart also functions as an endocrine organ: Forssmann, W. G., Hock, D., Mutt, V., et al. (1983). "The right auricle of the heart is an endocrine organ." *Anatomy and Embryology*, 168(3), 307–313; and Ogawa, T. and de Bold, A. J. (2014). "The heart as an endocrine organ." *Endocrine Connections*, 3(2), R31–R44.

p. 120 *cardiac natriuretic peptide*: Cantin, M. and Genest, J. (1986), "The heart as an endocrine gland." *Clinical and Investigative Medicine*, 9(4), 319–327.

p. 120 recent evidence indicates: De Dreu, C. K. and Kret, M. E. (2016). "Oxytocin conditions intergroup relations through upregulated in-group empathy, cooperation, conformity, and defense." *Biological Psychiatry*, 79(3), 165–173.

p. 120 sensitive magnetometers: Wakeup-world.com (2012). Retrieved from https://wakeup-world.com/2012/02/29/hearts-have-their-own-brain-and-consciousness/.

p. 120 spectral analysis: McCraty, R. and Tomasino, D. (2006). "The coherent heart: Heart-brain interactions, psychophysiological coherence, and the emergence of system wide order." (Publication No. 06- 022). Boulder Creek, Calif.: HeartMath Research Center, Institute of HeartMath.

p. 121 study of twenty-two heterosexual couples: Goldstein, P., Weissman-Fogel, I., Dumas, G., and Shamay-Tsoory, S. G. (2018). "Brain-to-brain coupling during handholding is associated with pain reduction." *Proceedings of the National Academy of Sciences*, 201703643.

p. 122 study from the Max Planck Institute in Berlin: Muller, V. and Lindenberger, U. (2011) "Cardiac and Respiratory Patterns Synchronize between Persons during Choir Singing." *PLoS One* 6(9), e24893.

p. 122 hearts that drum together: Gordon, I., Gilboa, A., Siegman, S., et al. (2020). "Physiological and Behavioral Synchrony predict Group cohesion and performance." *Scientific Reports*, 10(1), 1–12.

p. 122 Kidnapping-and-Cross-Fostering Study: Zayed, Amro and Robinson, Gene E. (2012). "Understanding the Relationship Between Brain Gene Expression and Social Behavior: Lessons from the Honey Bee." *Annual Review of Genetetics*, 46, 591–615.

p. 123 people younger than sixty-five: Madsen, N. L., et al. (2018). "Risk of Dementia in Adults With Congenital Heart Disease: Population-Based Cohort Study." *Circulation*, CIRCULATIONAHA-117.

p. 123 threefold increased risk of death: Glassman, A. H., Bigger, J. T., and Gaffney, M. (2007). "Heart Rate Variability in Acute Coronary Syndrome Patients with Major Depression, influence of Sertraline and Mood Improvement." *Archives of General Psychiatry*, 64, 9

p. 123 people who recently lost a spouse: Chirinos, D. A., Ong, J. C., Garcini, L. M., Alvarado, D., and Fagundes, C. (2019). "Bereavement, self-reported sleep disturbances, and inflammation: Results from Project HEART." *Psychosomatic Medicine*, 81(1), 67–73.

p. 123 schizophrenia is coronary artery disease: Ignaszewski, M. J., Yip, A., and Fitzpatrick, S. (2015). "Schizophrenia and coronary artery disease." *British Columbia Medical Journal*, 57(4), 154–157.

p. 124 At the 2018 Congress of the European Psychiatric Association: Kugathasan, P., Johansen, M. B., Jensen, M. B., Aagaard, J., and Jensen, S. E. (2018). "Coronary artery calcification in patients diagnosed with severe mental illness." *Circulation, Cardiovascular Imaging*, doi: 10.1161.

p. 124 acute myocardial infarction: Powers, Jenny (2018). "Increased Risk of Mortality From Heart Disease in Patients With Schizophrenia." Presented at EPA. https://www.firstwordpharma.com/node/1547343.

p. 124 morbidity and mortality in schizophrenic patients: Hennekens, C. H., Hennekens, A. R., Hollar, D., and Casey, D. E. (2005). "Schizophrenia and increased risks of cardiovascular disease." *American Heart Journal*, 150(6), 1115–1121; and Jindal, R., MacKenzie, E. M., Baker, G. B., and Yeragani, V. K. (2005). "Cardiac risk and schizophrenia." *Journal of Psychiatry and Neuroscience*, 30(6), 393.

p. 124 cardiac changes seen in these patients: Yeragani, V. K., Nadella, R., Hinze, B., Yeragani, S., and Jampala, V. C. (2000). "Nonlinear measures of heart period variability: Decreased measures of symbolic dynamics in patients with panic disorder." *Depression and Anxiety*, 12, 67–77; and Yeragani, V. K., Rao, K. A., Smitha, M. R., Pohl, R. B., Balon, R., and Srinivasan, K. (2002). "Diminished chaos of heart rate time series in patients with major depression." *Biological Psychiatry*, 51, 733–44.

p. 124 According to Laura Kubzansky of the Harvard School of Public Health: Laura Kubzansky in Rimer, S. and Drexler, M. (2011). "The biology of emotion and what it may teach us about helping people to live longer." *Harvard Public Health Review*, Winter, 813. Retrieved from https://www.hsph.harvard.edu/news/magazine/happiness-stress-heart-disease/.

p. 125 optimism cuts the risk of coronary heart disease by half: Kubzansky, L. D. and Thurston, R. C. (2007). "Emotional vitality and incident coronary heart disease: benefits of healthy psychological functioning." *Archives of General Psychiatry*, 64(12), 1393–1401.

p. 125 cerebral cortex: Armour, J. and Ardell, J. *Neurocardiology*. (New York: Oxford University Press, 1994).

p. 125 perception and emotional processing: Sandman, C. A., Walker, B. B., and Berka, C. (1982). "Influence of afferent cardiovascular feedback on behavior and the cortical evoked potential." *Perspectives in Cardiovascular Medicine;* and Van der Molen, M. W., Somsen, R.J.M. and Orlebeke, J. F. (1985). "The rhythm of the heart beat in information processing." *Advances in Psychophysiology*, 1, 1–88; and McCraty, R. "Heart–brain Neurodynamics: The Making of Emotions" (Publication No. 03–015). HeartMath Research Center, 2003.

p. 125 studies by David Glanzman: Chen, Shanping, Cai, Diancai, Glanzman, David L., et al. (2014). "Reinstatement of long-term memory following erasure of its behavioral and synaptic expression in Aplysia." *eLife*, 3, e03896.

p. 125 injecting them into the body of untrained snail recipients: Bédécarrats, A., Chen, S., Glanzman, D. L., et al. (2018). "RNA from trained Aplysia can induce an epigenetic engram for long-term sensitization in untrained Aplysia." *eNeuro*, 5(3); and Pearce, K., Cai, D., Roberts, A. C., et al. (2017). "Role of protein synthesis and DNA methylation in the consolidation and maintenance of long-term memory in Aplysia." *eLife*, 6.

p. 125 Shelley L. Berger at the University of Pennsylvania: Shelley L. Berger in public release: University of Pennsylvania (May 31, 2017). "Metabolic enzyme fuels molecular machinery of memory." *Penn Medicine News*. Penn study finds epigenetics key to laying down spatial memories in mouse brain, providing possible new neurological medications, https://www.pennmedicine.org/news/news-releases/2017/may/metabolic-enzyme-fuels-molecular-machinery-of-memory.

p. 126 Shomrat and Levin's research on planaria: Shomrat, T. and Levin, M. (2013). "An automated training paradigm reveals long-term memory in planarians and its persistence through head regeneration." *Journal of Experimental Biology*, 216(20), 3799–3810.

p. 126 heart transplant patients at University Hospital in Vienna: Bunzel, B., Schmidl-Mohl, B., Grundböck, A., and Wollenek, G. (1992). "Does changing the heart mean changing personality? A retrospective inquiry on 47 heart transplant patients." *Quality of Life Research*, 1(4), 251–256.

p. 126 study of heart transplants in Israel: Inspector, Y., Kutz, I., and Daniel, D. (2004). "Another person's heart: Magical and rational thinking in the psychological adaptation to heart transplantation." *The Israel Journal of Psychiatry and Related Sciences*, 41(3), 161.

p. 127 reviewing the literature on heart transplants: Liester, M. and Liester, M. (2019). "Personality changes following heart transplants: can epigenetics explain these transformations?" *Medical Hypotheses*, 135(3), 0.1016/j.mehy.2019.109468; and Liester, Mitchell and Liester, Maya. (2019). "A Retrospective Phenomenological Review of Acquired Personality Traits Following Heart Transplantation." April 2019. Conference: 27th European Congress of Psychiatry, Warsaw.

p. 127 ten cases of heart or heart-lung transplants: Pearsall, Paul, et al. "Organ transplants and cellular memories." *Nexus Magazine*, April/May 2005, 12, 3.

p. 127 recipients experienced profound changes: Pearsall, P., Schwartz, G. E., and Russek, L. G. (2002). "Changes in heart transplant recipients that parallel the personalities of their donors." *Journal of Near-Death Studies*, 20(3), 191–206.

p. 129 memory exists in every heart: Pearsall, P., Schwartz, G. E., and Russek, L. G. (2002). Changes in heart transplant recipients that parallel the personalities of their donors. *Journal of Near-Death Studies*, 20(3), 191–206.

p. 129 best known account is that of Claire Sylvia: Sylvia, Claire *A Change of Heart*. (New York: Warner Books, 1998).

p. 129 "The idea . . . is unimaginable.": John Schroeder in Skeptic's Dictionary Online: http://skepdic.com/cellular.html.

p. 130 "It's been out in the pop culture . . .": Heather Ross in De Giorgio Lorriana (March 28, 2012). "Can a heart transplant change your personality?" *Toronto Star*. Retrieved from https://www.thestar.com/news/world/2012/03/28/can_a_heart_transplant_change_your_personality.html#:~:text=Dr.,that%20such%20a%20thing%20happens.

p. 130 explanations entertained by skeptics: Lunde, D. T. (1967). "Psychiatric complications of heart transplants." *American Journal of Psychiatry*, 124:1190–1195; Mai, F. M. (1986). "Graft and donor denial in heart transplant recipients." *American Journal of Psychiatry*; 143, 1159–1161 and Kuhn, W. F., et al. (1988). "Psychopathology in heart transplant candidates." *The Journal of Heart Transplantation*, 7, 223–226.

p. 130 researchers at the University of Bordeaux: Wagner, M., Helmer, C., Samieri, C., et al. (2018). "Evaluation of the concurrent trajectories of cardiometabolic risk factors in the 14 years before dementia." *JAMA Psychiatry*, *75*(10), 1033–1042.

p. 130 recommendation to eat fish: Rimm, E. B., Appel, L. J., Lichtenstein, A. H., et al. (2018). "Seafood long-chain n-3 polyunsaturated fatty acids and cardiovascular disease: A science advisory from the American Heart Association." *Circulation*, 138(1), e35–e47.

p. 130 individuals who have close and meaningful relationships: Cohen, S., Syme, S., eds. *Social Support and Health*. (Orlando, Fla.: Academic Press, 1985); Uchino, B. N., Cacioppo, J. T., Kiecolt-Glaser, J. K. (1996). "The relationship between social support and physiological processes: a review with emphasis on underlying mechanisms and implications for health." *Psychological Bulletin*, 119(3), 488–531. doi:10.1037/0033-2909.119.3.488; and Ornish, D. *Love and Survival: The Scientific Basis for the Healing Power of Intimacy*. (New York: HarperCollins, 1998).

p. 131 psychophysiological coherence: McCraty, R. (2000), Psychophysiological coherence: A link between positive emotions, stress reduction, performance, and health. Proceedings of the Eleventh International Congress on Stress, Mauna Lani Bay, Hawaii.

p. 132 electric stimulation: Chua, E. F. and Ahmed, R. (2016). "Electrical stimulation of the dorsolateral prefrontal cortex improves memory monitoring." *Neuropsychologia*, 85, 74–79.

9. The Brain-Mind Conundrum and the Rise of Quantum Biology

p. 136 philosopher David Chalmers: Chalmers, David (1995). "Facing Up to the Problem of Consciousness." *Journal of Consciousness Studies*, 2(3), 200–219.

p. 136 philosopher Owen Flanagan: Flanagan, Owen, *The Science of the Mind.* 2nd ed. (Cambridge, Mass.: MIT Press, 1991).

p. 137 conscious mind is like an interpreter: Morsella, E., Godwin, C. A., Gazzaley, A., et al. (2016). "Homing in on consciousness in the nervous system: An action-based synthesis." *Behavioral and Brain Sciences*, 39.

p. 137 Pierre Jean Georges Cabanis: Blackburn, Simon (2008). *The Oxford Dictionary of Philosophy* (2nd rev. ed.). (Oxford, UK: Oxford University Press, 2016).

p. 137 Dan Dennett's term *epiphenomenalism*: Dennett, D. C. *Consciousness Explained*, (Boston: Little, Brown, 1991).

p. 137 consciousness was thought to reside in the prefrontal cortex: Pal, D., Dean, J. G., Hudetz, A. G., et al. (2018). "Differential role of prefrontal and parietal cortices in controlling level of consciousness." *Current Biology*, 28, 2145.e5–2152.e5. doi: 10.1016/j.cub.2018.05.025.

p. 137 Some authors: Boly, M., Massimini, M., Tsuchiya, N., Postle, B. R., Koch, C., and Tononi, G. (2017). "Are the neural correlates of consciousness in the front or in the back of the cerebral cortex? Clinical and neuroimaging evidence." *Journal of Neuroscience*, 37, 9603–9613.

p. 137 consciousness likely occurs in dendrites: Eccles, J. C. (1992). "Evolution of consciousness." *Proceedings of the National Academy of Sciences*, 897320–7324.

p. 137 Karl Pribram: Pribram, Karl (2012). *Quantum Implications: Essays in Honour of David Bohm*; chapter "The Implicate Brain" (New York: Routledge, 1987).

p. 138 *The Cambridge Declaration on Consciousness:* http://worldanimal.net/images/stories/documents/Cambridge-Declaration-on-Consciousness.pdf.

p. 138 Anesthetic agents do not suppress brain function globally: Bonhomme, V., Staquet, C., Vanhaudenhuyse, A., Gosseries, O., et al. (2019) "General Anesthesia: A Probe to Explore Consciousness." *Frontiers in Systems Neuroscience*, 13, 36.

p. 138 whether consciousness is fully lost during anesthesia: Scheinin, A., Kallionpää, R. E., Revonsuo, A., et al. (2018). "Differentiating drug-related and state-related effects of dexmedetomidine and propofol on the electroencephalogram." *Anesthesiology: The Journal of the American Society of Anesthesiologists*, 129(1), 22–36.

p. 139 Conscious selfhood: Seth, Anil K. (2017). "The real problem: It looks like scientists and philosophers might have made consciousness far more mysterious than it needs to be." *Aeon Magazine.*

p. 139 brain activity preceded their actions: Libet, B. (1985). "Unconscious cerebral initiative and the role of conscious will in voluntary action." *Behavioral and Brain Sciences*, 8(4), 529–539.

p. 140 interpretation of these studies: Saigle, Victoria, Dubljević, Veljko, and Racine, Eric (2018). "The Impact of a Landmark Neuroscience Study on Free Will: A Qualitative Analysis of Articles Using Libet and Colleagues' Methods." *AJOB Neuroscience*, 9(1), 29.

p. 140 functional neuroanatomy of volitional behavior: Brass, M. and Haggard, P. (2008). "The what, when, whether model of intentional action." *The Neuroscientist*, 14(4), 319–325.

p. 140 Sir Roger Penrose: Penrose, R. *The Emperor's New Mind.* (Oxford, UK: Oxford University Press, 1989); and Penrose, R. *Shadows of the Mind: A Search for the Missing Science of Consciousness.* (Oxford, UK: Oxford University Press, 1994).

p. 140 *quantum mechanism*: Penrose, R. *The Emperor's New Mind*. (Oxford, UK: Oxford University Press, 1989).

p. 140 Libet's backward time effects: Bierman, D. J. (2001). "New developments in presentiment research, or the nature of time." Presentation to the Bial Foundation Symposium, Porto, Portugal; and Radin, D. I., Taft, R., and Yount, G, (2004). "Possible effects of healing intention on cell cultures and truly random events." *Journal of Alternative and Complementary Medicine,* 10, 103–112.

p. 140 Dr. Masaru Emoto: Emoto, Masaru. *The Hidden Messages in Water.* (New York: Atria Books, 2005).

p. 141 birth memories: Chamberlain, David. *Babies Remember Birth.* (Los Angeles: Jeremy P. Tarcher, 1988).

p. 142 James Braid: Hergenhahn, B. and Henley, T. *An Introduction to the History of Psychology.* (Andover, UK: Cengage Learning, 2013).

p. 142 skeptical of hypnosis: Jensen, M. P., Jamieson, G. A., Santarcangelo, E. L., Terhune, D. B., et al. (2017). "New directions in hypnosis research: strategies for advancing the cognitive and clinical neuroscience of hypnosis." *Neuroscience of Consciousness*, 2017(1), nix004.

p. 142 hypnosis should be included: Holdevici, I. (2014). "A brief introduction to the history and clinical use of hypnosis." *Romanian Journal of Cognitive Behavioral Therapy and Hypnosis,* 1(1), 1–5.

p. 143 color of pills made a difference: De Craen, A. J., Roos, P. J., De Vries, A. L., and Kleijnen, J. (1996). "Effect of colour of drugs: systematic review of perceived effect of drugs and of their effectiveness." *BMJ*, 313(7072), 1624–1626.

p. 144 team of neuroscientists of TU Dresden: Tabas, A., Mihai, G., Kiebel, S., Trampel, R., and von Kriegstein, K. (2020). "Abstract rules drive adaptation in the subcortical sensory pathway." *eLife*, 9, e64501.

p. 145 Sam Parnia: Parnia, S., Spearpoint, K., Wood, M., et al. (2014). "AWARE—AWAreness during REsuscitation—A prospective study." *Resuscitation*, 85(12), 1799–1805.

p. 146 special psychic connection: Playfair, Guy Lyon. *Twin Telepathy: The Psychic Connection.* (New York: Vega Books, 2003)

p. 146 past life memories of children: Stevenson, Ian, (1984). "American Children Who Claim To Remember Previous Lives." *Journal of Nervous and Mental Disease*, 171, 742–748.

p. 147 *microtubules*: Hameroff, Stuart (2014). "Consciousness, Microtubules, and 'Orch OR': A 'Space-time Odyssey." *Journal of Consciousness Studies*, 21(3–4), 126–158; and Volk, Steve (2018). "Down The Quantum Rabbit Hole." *Discover.*

p. 147 single-celled paramecium: Volk, Steve (2018). "Down The Quantum Rabbit Hole." *Discover.*

p. 148 quantum physics might play a role in consciousness: Volk, Steve (2018). "Down The Quantum Rabbit Hole." *Discover.*

p. 148 Consciousness is more like music: Hameroff, Stuart (November 2015). "Is your brain really a computer, or is it a quantum orchestra tuned to the universe?" *Interalia Magazine.* Retrieved from https://www.interaliamag.org/articles/stuart-hameroff-is-your-brain-really-a-computer-or-is-it-a-quantum-orchestra-tuned-to-the-universe/.

p. 148 quantum coherence: Elsevier (2014). "Discovery of quantum vibrations in 'microtubules' corroborates theory of consciousness." *Physics of Life Reviews.*

p. 148 quantum effects can happen in biological systems: Lambert, N., Chen, Y. N., Cheng, Y. C., Li, C. M., Chen, G. Y., and Nori, F. (2013). "Quantum biology." *Nature Physics*, 9(1), 10.

p. 148 Anthony Hudetz: A. Hudetz in Volk, Steve (2018). "Down The Quantum Rabbit Hole." *Discover.*

p. 148 Post-Materialist Science: Open Sciences, http://opensciences.org/about/manifesto-for-a-post-materialist-science.

p. 150 the physical world is no longer the primary or sole component of reality: (Open Science, Manifesto for a Post-Materialist Science). Open Sciences, http://opensciences.org/about/manifesto-for-a-post-materialist-science.

p. 151 *"quantum leaps"*: Schrödinger, E. *What Is Life? With Mind and Matter and Autobiographical Sketches.* Schrödinger, E. (Cambridge, UK: Cambridge University Press, 1967).

p. 151 *"Quantum Biology"*: Lowdin, P. O. (1965). "Quantum genetics and the aperiodic solid. Some aspects on the Biological problems of heredity, mutations, aging and tumors in view of the quantum theory of the DNA molecule." *Advances in Quantum Chemistry*, 2, 213–360.

p. 151 both states remain possible outcomes of a measurement: Ball, Philip (2017). "Quantum common sense: Despite its confounding reputation, quantum mechanics both guides and helps explain human intuition." *Aeon Magazine.*

p. 152 lucid dreaming as an integral part of therapy: Collier, Sandra. *Wake Up to Your Dreams.* (Toronto: Scholastic Canada, 1996).

p. 152 *Three Faces of Eve*: Thigpen, C. H. and Cleckley, H. M. *The 3 Faces of Eve.* (New York: McGraw-Hill, 1957).

p. 152 distinct personality states: American Psychiatric Association. (2013). *Diagnostic and statistical manual of mental disorders (DSM-5).* (Washington: American Psychiatric Publishing, 2013).

p. 153 *quantum teleportation*: "Double quantum-teleportation milestone is Physics World 2015 Breakthrough of the Year physicsworld.com." *Physicsworld.com.*

p. 153 accepted theory of olfaction: Sell, C. S. (2006). "On the unpredictability of odor." *Angewandte Chemie International Edition*, 45, 6254–6261.

p. 153 vibration theory: Turin, Luca. *The Secret of Scent: Adventures in Perfume and the Science of Smell.* (New York: Ecco, 2006).

p. 153 each molecule's pattern of vibration: Schillinger, Liesl (2003). "Odorama." *New York Times.*

p. 153 olfactory receptors in humans: Olender, T., Lancet, D., Nebert, D. W. (2008–2009). "Update on the olfactory receptor (*OR*) gene superfamily." *Human Genomics*, 3, 87–97.

p. 153 "swipe card" model: Brookes, J. C., Horsfield, A. P., and Stoneham, A. M. (2012). "The swipe card model of odorant recognition." *Sensors*, 12(11), 15709–15749.

p. 154 Turin's proposal of vibration frequencies: Brookes, J. C., Hartoutsiou, F., Horsfield, A. P., and Stoneham, A. M. (2007). "Could humans recognize odor by phonon assisted tunneling?" *Physical Review Letters*, 98(3), 038101.

p. 154 founding patent on the *Poised Realm*: Kauffman, S., Niiranen, S., and Vattay, G. (2014). *U.S. Patent No. 8,849,580.* Washington: U.S. Patent and Trademark Office.

p. 154 where consciousness reigns: Kauffman, S. (2010). "Is There A 'Poised Realm' Between the Quantum and Classical Worlds?" *Cosmos and Culture.*

p. 155 view proposed by the philosopher Albert North Whitehead: Whitehead, A. N. *Process and Reality.* (New York: MacMillan, 1929); and Whitehead, A. N. *Adventures of Ideas.* (London: MacMillan, 1933).

p. 155 Kauffman hypothesis: Kauffman, S. (2010). "Is There A 'Poised Realm' Between the Quantum and Classical Worlds?" *Cosmos and Culture.*; and Kauffman, S. *Humanity in a Creative Universe.* (Oxford, UK: Oxford University Press, 2016).

p. 156 telepathy/sympathetic communication and telekinesis: Radin, D. *Entangled Minds*, (New York: Pocket Books, 2006); and Kauffman, S. (2010). "Is There A 'Poised Realm' Between the Quantum and Classical Worlds?" *Cosmos and Culture*; and Caswell, J., Dotta, B., and Persinger, M. (2014). "Cerebral Biophoton Emission as a Potential Factor in Non-Local Human-Machine Interaction." *NeuroQuantology*, 12(1), 1–11.

p. 157 how the mind affects matter: Hameroff, S. and Chopra, D. (2012) "The 'quantum soul': A scientific hypothesis," in Moreira-Almeida, A. and Santos, F. S. (eds.) *Exploring Frontiers of the Mind–Brain Relationship*, 79–93, (New York: Springer, 2011).

p. 157 precursor of the mind: Frank, Adam (2017). "Minding Matter." *Aeon Magazine*

10. Significance

p. 160 researchers from the German Center for Neurodegenerative Diseases: Islam, M. R., Lbik, D., Sakib, M. S., Maximilian Hofmann, R., Berulava, T., Jiménez Mausbach, M., and Fischer, A. (2021). "Epigenetic gene expression links heart failure to memory impairment." *EMBO Molecular Medicine*, 13(3), e11900.

p. 161 linked the gut microbiome to social behavior: Nguyen, T. T., Zhang, X., Wu, T. C., Liu, J., Le, C., Tu, X. M., and Jeste, D. V. (2021). "Association of Loneliness and Wisdom with Gut Microbial Diversity and Composition: An Exploratory Study." *Frontiers in Psychiatry*, 12, 395.

p. 162 signals that muscles send when stressed: Rai, Mamta, Coleman, Zane, Demontis, Fabio, et al. (2021). "Proteasome stress in skeletal muscle mounts a long-range protective response that delays retinal and brain aging." *Cell Metabolism*,

p. 162 James Herman: Herman, James (September 12, 2019). In "Fight or Flight May Be in Our Bones" by Diana Kwon. *Scientific American*.

p. 164 fifteen tiny organelles and forty-two million protein molecules: Ho, B., Baryshnikova, A., and Brown, G. W. (2018). "Unification of protein abundance datasets yields a quantitative Saccharomyces cerevisiae proteome." *Cell Systems*, 6(2), 192–205.

p. 165 quorum sensing: Wong, G. C., Antani, J. D., Bassler, Bonnie, Dunkel, J., et al. (2021). "Roadmap on emerging concepts in the physical biology of bacterial biofilms: from surface sensing to community formation." *Physical Biology*.

p. 165 Gürol Süel . . . bacteria in biofilms: Ang, C. Y., Bialecka-Fornal, M., Weatherwax, C., Larkin, J. W., Prindle, A., Liu, J., and Süel, G. M. (2020). "Encoding membrane-potential-based memory within a microbial community." *Cell Systems*, 10(5), 417–423; and Popkin, Gabriel, (2017). "Bacteria Use Bursts of Electricity to Communicate." *Quanta Magazine*.

p. 165 James Shapiro: Shapiro, J. A. (2007). "Bacteria are small but not stupid: cognition, natural genetic engineering and socio-bacteriology." *Studies in History and Philosophy of Science Part C: Studies in History and Philosophy of Biological and Biomedical Sciences*, 38(4), 807–819.

p. 169 A study at Charité-University Medicine Berlin on telomeres: Verner, G., Epel, E., and Entringer, S. (2021). "Maternal psychological resilience during pregnancy and newborn telomere length: A prospective study." *American Journal of Psychiatry*, 178, 2.

p. 170 a study from Oregon State University: Turner, S. G. and Hooker, K. (2020). "Are Thoughts About the Future Associated with Perceptions in the Present? Optimism, Possible Selves, and Self-Perceptions of Aging." *The International Journal of Aging and Human Development*, 0091415020981883.

p. 171 Researchers from Finland's University of Turku: Tuominen, J., Kallio, S., Kaasinen, V., and Railo, H. (2021). "Segregated brain state during hypnosis." *Neuroscience of Consciousness*, 2021(1), niab002.

p. 172 Ted Kaptchuk of Beth Israel Deaconess Medical Center: Ted Kaptchuk in "The Power of the Placebo Effect." https://www.health.harvard.edu/mental-health/the-power-of-the-placebo-effect#:~:text=%22The%20plac.

p. 173 scientists at Imperial College London: Rodgers, Paul (2014). "Einstein Was Right: You Can Turn Energy Into Matter." https://www.forbes.com/sites/paulrodgers/2014/05/19/einstein-was-right-you-can-turn-energy-into-matter/?sh=2c58c85126ac.

BIBLIOGRAPHY

Abdo, H., Calvo-Enrique, L., and Ernfors, P. (2019). "Specialized cutaneous Schwann cells initiate pain sensation." *Science*, 365(6454), 695–699.

Adamatzky, A. I. (2014). "Route 20, autobahn 7, and slime mold: approximating the longest roads in USA and Germany with slime mold on 3D terrains." *IEEE Transactions on Cybernetics*, 44(1), 126–136.

Adamatzky, A. and Jones, J. (2010). "Road planning with slime mould: If Physarum built motorways it would route M6/M74 through Newcastle." *International Journal of Bifurcation and Chaos*, 20(10), 3065–3084.

Albrecht-Buehler, G. (1985). "Is cytoplasm intelligent too?" In *Cell and Muscle Motility* (1–21), (Dordrecht, Netherlands: Springer, 1981).

Alcino Silva in Han, X., Chen, M., Wang, F., Windrem, M., Wang, S., Shanz, S., and Silva, A. J. (2013). "Forebrain engraftment by human glial progenitor cells enhances synaptic plasticity and learning in adult mice." *Cell Stem Cell*, 12(3), 342–353. 8.

Allen, J. M., Mailing, L. J., Woods, J. A., et al. (2018). "Exercise alters gut microbiota composition and function in lean and obese humans." *Medicine and Science in Sports and Exercise*, 50(4), 747–757.

Allen, K., Fuchs, E. C., Jaschonek, H., Bannerman, D. M., and Monyer, H. (2011). "Gap Junctions between Interneurons Are Required for Normal Spatial Coding in the Hippocampus and Short-Term Spatial Memory." *Journal of Neuroscience*, 31(17): 6542–6552.

Allman, J. M. (New York: Scientific American Library, 1999).

Almgren, Malin, Schlinzig, Titus, Gomez-Cabrero, David, Ekström, Tomas J., et al. (2014). "Cesarean section and hematopoietic stem cell epigenetics in the newborn infant—implications for future health?" *American Journal of Obstetrics and Gynecology*, 22(5), 502.e1–502.e8.

Alvarez, L., Friedrich, B. M., Gompper, G. and Kaupp, U. B. (2014). "The computational sperm cell." *Trends in Cell Biology*, 24, 198–207.

AMA adopts new policies at annual meeting [news release]. Chicago: American Medical Association; June 20, 2011. http://www.Ama-assn.org/ama/pub/news/news/2011-new-policies-adopted.page. Accessed December 6, 2011.

American Academy of Pediatrics (AAP). (2012). "Breastfeeding and the use of human milk." *Pediatrics*, 129, e827–e841. http://pediatrics.aappublications.org/content/129/3/e827.full.pdf+html.

American College of Obstetricians and Gynecologists Committee on Obstetric Practice. Committee opinion no. 637: marijuana use during pregnancy and lactation. *American Journal of Obstetrics and Gynecology*, 2015, 126(1), 234–238.

American Psychiatric Association. *Diagnostic and statistical manual of mental disorders (DSM-5).* (Washington: American Psychiatric Publishing, 2013).

American Psychiatric Association. *Diagnostic and statistical manual of mental disorders* (3rd ed.). (Washington: American Psychiatric Publishing, 1987).

American Psychiatric Association. *Diagnostic and statistical manual of mental disorders* (4th edition), (Washington: American Psychiatric Press, 2000).

Andaloussi, S. E., Mäger, I., Breakefield, X. O., and Wood, M. J. (2013). "Extracellular vesicles: biology and emerging therapeutic opportunities." *Nature Reviews Drug Discovery*, 12(5), 347–357.

answers.yahoo.com/question/index?quid=20071123141714AAzXSar.

Anthony Hudetz and Robert Pearce (eds.), *Suppressing the Mind: Anesthetic Modulation of Memory and Consciousness.* Contemporary Clinical Neuroscience (New York: Humana Press, 2010)

Appelfeld, A. (2005). *Geschichte eines Lebens.* (Berlin: Rowohlt, 2005).

Archaea: In *Wikipedia*: https://en.wikipedia.org/wiki/Archaea.

Armitage, R., Flynn, H., Marcus, S., et al. (2005). "Early developmental changes in sleep in infants: The impact of maternal depression." *Sleep,* 32(5), 693–696.

Armour, J. and Ardell, J. *Neurocardiology.* (New York: Oxford University Press, 1994).

Armour, J., "Anatomy and function of the intrathoracic neurons regulating the mammalian heart." *Reflex Control of the Circulation*, I. H. Zucker and J. P. Gilmore, eds. (Boca Raton, Fla.: CRC Press, 1991), 1–37.

Armour, J. A. (2008). "Potential clinical relevance of the 'little brain' on the mammalian heart." *Experimental Physiology*, 93(2), 165–176.

Ang, C. Y., Bialecka-Fornal, M., Weatherwax, C., Larkin, J. W., Prindle, A., Liu, J., and Süel, G. M. (2020). "Encoding membrane-potential-based memory within a microbial community." *Cell Systems*, 10(5), 417–423.

Arntz, W. (producer) and Arntz, W. (director). (2004) *What the Bleep Do We Know!?* [motion picture]. United States: Samuel Goldwyn Films.

Association for Pre- and Perinatal Psychology and Health (APPPAH) birthpsychology.com.

Ataide, Marco A., Komander, Karl, Kastenmüller, Wolfgang, et al. (2020). "BATF3 programs CD8 T cell memory." *Nature Immunology*, 1–11.

Atom. In *Wikipedia*. Retrieved from https://en.wikipedia.org/wiki/Atom.

Axness, M. (2016). "To Be or Not to Be: Prenatal Origins of the Existential 'Yes' v. the Self Struggle." *Journal of Prenatal & Perinatal Psychology & Health*, 31(2).

Bachiller, S., Jiménez-Ferrer, I., Boza-Serrano, A., et al. (2018). "Microglia in neurological diseases: a road map to brain-disease dependent-inflammatory response." *Frontiers in Cellular Neuroscience*, 12, 488.

Bacteria: In *Wikipedia*. https://en.wikipedia.org/wiki/Bacteria.

Bagge, C. N., Henderson, V. W., Laursen, H. B., Adelborg, K., Olsen, M., and Madsen, N. L. (2018). "Risk of Dementia in Adults With Congenital Heart Disease: Population-Based Cohort Study." *Circulation*, CIRCULATIONAHA-117.

Baldscientist (2011). Playing With Worms. Retrieved from https://baldscientist.wordpress .com/2011/09/23/playing-with-wormies/

Ball, Philip (2017). "Quantum common sense: Despite its confounding reputation, quantum mechanics both guides and helps explain human intuition." *Aeon Magazine.*

Baluška, F. and Mancuso, S. (2009). "Deep evolutionary origins of neurobiology: Turning the essence of 'neural' upside-down." *Communicative & Integrative Biology*, 2(1), 60–65.

Baluška, F. and Levin, M. (2016). "On having no head: cognition throughout biological systems." *Frontiers in Psychology*, 7, 902.

Barocas, H. A. and Barocas, C. B. (1979). "Wounds of the fathers: The next generation of Holocaust victims." *International Review of Psycho-Analysis.*

Barone, V., Lang, M., Heisenberg, C. P., et al. (2017). "An effective feedback loop between cell-cell contact duration and morphogen signaling determines cell fate." *Developmental Cell*, 43(2), 198–211.

Battaglia, D., Veggiotti, P., Lettori, D., Tamburrini, G., Tartaglione, T., Graziano, A., and Colosimo, C. (2009). "Functional hemispherectomy in children with epilepsy and CSWS due to unilateral early brain injury including thalamus: sudden recovery of CSWS." *Epilepsy Research*, 87(2–3), 290–298.

Bayan, L., Koulivand, P. H. and Gorji, A. (2014). "Garlic: a review of potential therapeutic effects." *Avicenna Journal of Phytomedicine*, 4(1), 1.

Beblo, Julienne. (Oct. 16, 2018). *"Are You Smarter Than an Octopus?"* National Marine Sanctuary. Retrieved from https://marinesanctuary.org/blog/are-you-smarter-than-an-octopus/?gclid=EAIaIQobChMIxMD9qZa04QIVhyaGCh2Zhgy8EA AYASAAEgKTGPD_BwE.

Bédécarrats, A., Chen, S., Glanzman, D. L., et al. (2018). "RNA from trained Aplysia can induce an epigenetic engram for long-term sensitization in untrained Aplysia." *eNeuro*, 5(3).

Benenson, J. F. (1993). "Greater preference among females than males for dyadic interaction in early childhood." *Child Development*, 64(2), 544–555.

Beth Israel Deaconess Medical Center (2016). "Insight into the seat of human consciousness." *ScienceDaily*.

Bevington, Sarah L. Cauchy, Cockerill, Peter N., et al. (2016). "Inducible chromatin priming is associated with the establishment of immunological memory in T cells." *The EMBO J*, https://pubmed.ncbi.nlm.nih.gov/26796577/.

Biggar K. K., Storey, K. B. (2011) The emerging roles of microRNAs in the molecular responses of metabolic rate depression. *Journal of Molecular Cell Biology* 3:167–175.

Bierman, D. J. (2001). New developments in presentiment research, or the nature of time. Presentation to the Bial Foundation Symposium, Porto, Portugal.

Bissiere, S., Zelikowsky, M., Fanselow, M. S., et al. (2011) "Electrical synapses control hippocampal contributions to fear learning and memory." *Science* 331(6013)87–91.

Blackburn, Simon (2008). *The Oxford Dictionary of Philosophy* (2nd rev. ed.). Oxford University Press.

Blackiston, D. J., Shomrat, T., and Levin, M. (2015). "The stability of memories during brain remodeling: a perspective." *Communicative & Integrative Biology*, 8(5), e1073424.

Blackiston, D. J., Silva, Casey E., and Weiss, M. R. (2008) "Retention of memory through metamorphosis: can a moth remember what it learned as a caterpillar?," *PLoS ONE* 3(3): e1736.

Blaser, Martin J. (2014). "The Way You're Born Can Mess with the Microbes You Need to Survive." www.wired.

Blatt, S. J., Zuroff, D. C., et al. (2010). "Predictors of sustained therapeutic change." *Psychotherapy Research*, 20(1), 37–54.

Boly, M., Massimini, M., Tsuchiya, N., Postle, B. R., Koch, C., and Tononi, G. (2017). "Are the neural correlates of consciousness in the front or in the back of the cerebral cortex? Clinical and neuroimaging evidence." *Journal of Neuroscience*, 37, 9603–9613.

Bonner, John. *Life Cycles: Reflections of an Evolutionary Biologist*. (Princeton, N.J.: Princeton University Press, 1992).

Bonner, John. *Sixty Years of Biology: Essays on Evolution and Development*. (Princeton, N.J.: Princeton University Press, 1996).

Bonner, John. *First Signals: The Evolution of Multicellular Development*. (Princeton, N.J.: Princeton University Press, 2000).

Bonner, John. *Lives of a Biologist: Adventures in a Century of Extraordinary Science*. (Cambridge, Mass.: Harvard University Press, 2002).

Bonner, John. *The Social Amoebae: The Biology of the Cellular Slime Molds*. (Princeton, N.J.: Princeton University Press, 2009).

Bonner, John. *Why Size Matters: From Bacteria to Blue Whales*. (Princeton, N.J.: Princeton University Press, 2006).

Borghol, N., Suderman, M., McArdle, W., Racine, A., Hallett, M., Pembrey, M., and Szyf, M. (2012). "Associations with early-life socio-economic position in adult DNA methylation." *International Journal of Epidemiology*, 41(1), 62–74.

Bormann, J. E. and Carrico, A. W. (2009). "Increases in Positive Reappraisal Coping During a Group-Based Mantram Intervention Mediate Sustained Reductions in Anger in HIV-Positive Persons." *International Journal of Behavioral Medicine* 16: 74. doi:10.1007/s12529-008-9007-3.

Botstein, D. and Risch, N. (2003). "Discovering genotypes underlying human phenotypes: Past successes for mendelian disease, future approaches for complex disease." *Nature Genetics*, 33, 228–237.

Branchi, I., Francia, N., and Alleva, E. (2004). "Epigenetic control of neurobehavioral plasticity: the role of neurotrophins." *Behavioral Pharmacology*, 15, 353–362. [PubMed: 15343058]

Brass, M. and Haggard, P. (2008). "The what, when, whether model of intentional action." *The Neuroscientist*, 14(4), 319–325.

Bräuer, J., Hanus, D., and Uomini, N. (2020). "Old and New Approaches to Animal Cognition: There Is Not 'One Cognition'." *Journal of Intelligence*, 8(3), 28.

Braun, J. M., Kalkbrenner, A. E., Calafat, A. M., Yolton, K., Ye, X., Dietrich, K. N., and Lanphear, B. P. (2011). "Impact of early-life bisphenol A exposure on behavior and executive function in children." *Pediatrics*, 128(5), 873–882.

Brave Heart, M.Y.H., Chase, J., Elkins, J., and Altschul, D. B. (2011). "Historical trauma among indigenous peoples of the Americas: Concepts, research, and clinical considerations." *Journal of Psychoactive Drugs*, 43(4), 282–290.

Bremner, J. D. (2003). "Long-term effects of childhood abuse on brain and neurobiology." *Child & Adolescent Psychiatric Clinics of North America*, 12, 271–292. [PubMed: 12725012]

Brenman, Patricia A., Mednick, Sarnoff A., et al. (2002). "Relationship of Maternal Smoking During Pregnancy With Criminal Arrest and Hospitalization for Substance Abuse in Male and Female Adult Offspring." *The American Journal of Psychiatry*, 59(1), 48–54.

Brezinka, C., Huter, O., Biebl, W., and Kinzl, J. (1994). 'Denial of Pregnancy: Obstetrical aspects." *Journal of Psychosomatic Obstetrics and Gynecology*, 15, 1–8.

Brinkworth, J. F. and Alvarado, A. S. (2020). "Cell-Autonomous Immunity and The Pathogen-Mediated Evolution of Humans: Or How Our Prokaryotic and Single-Celled Origins Affect The Human Evolutionary Story." *The Quarterly Review of Biology*, 95(3), 215–246.

Britannica. https://www.britannica.com/science/morphogenesis.

Brookes, J. C., Hartoutsiou, F., Horsfield, A. P., and Stoneham, A. M. (2007). "Could humans recognize odor by phonon assisted tunneling?" *Physical Review Letters*, 98(3), 038101.

Brookes, J. C., Horsfield, A. P., and Stoneham, A. M. (2012). "The swipe card model of odorant recognition." *Sensors*, 12(11), 15709–15749.

Brown, D. G., Soto, R., Fujinami, R. S., et al. (2019). "The microbiota protects from viral-induced neurologic damage through microglia-intrinsic TLR signaling." *eLife*, 8, e47117.

Bruel-Jungerman, E., Davis, S., and Laroche, S. (2007). "Brain plasticity mechanisms and memory: a party of four." *The Neuroscientist*, 13, 492–505. doi: 10.1177/1073858407302725.

Bu, L., Jiang, X., Martin-Puig, S., Caron, L., Zhu, S., Shao, Y., and Chien, K. R. (2009). "Human ISL1 heart progenitors generate diverse multipotent cardiovascular cell lineages." *Nature*, 460(7251), 113.

Buchanan, Tony W. (2007). "Retrieval of Emotional Memories." *Psychological Bulletin*, 61–779. doi: 10.1037/0033-2909.133.5.761.

Buka, S. L., Shenassa, E. D., and Niaura, R. (2003). "Elevated risk of tobacco dependence among offspring of mothers who smoked during pregnancy: A 30-year prospective study." *American Journal of Psychiatry*, 160(11), 1978–1984.

Bunzel, B., Schmidl-Mohl, B., Grundböck, A., and Wollenek, G. (1992). "Does changing the heart mean changing personality? A retrospective inquiry on 47 heart transplant patients." *Quality of Life Research*, 1(4), 251–256.

Burkett, J. P., Andari, E., Young L. J., et al. (2016). "Oxytocin-dependent consolation behavior in rodents." *Science*, 351(6271), 375–378. doi: 10.1126/science.aac4785.

Burton, N. O., Furuta, T., Webster, A. K., Kaplan, R. E., Baugh, L. R., Arur, S., and Horvitz, H. R. (2017). "Insulin-like signaling to the maternal germline controls progeny response to osmotic stress." *Nature Cell Biology*, 19(3), 252–257.

Buss, C., Entringer, S., Swanson, J. M., and Wadhwa, P. D. (2012). "The Role of Stress in Brain Development: The Gestational Environment's Long-Term Effects on the Brain." *Cerebrum: the Dana Forum on Brain Science*, 4.

Buss, Claudia, Davis, Elysia Poggi, Sandman, Curt A., et al. (2010). "High pregnancy anxiety during mid-gestation is associated with decreased gray matter density in 6–9 year-old children." *Psychoneuroendocrinology*, 35(1), 141–153.

Bustan, Muhammad N. and Coker, Ann L. (1994). "Maternal Attitude toward Pregnancy and the Risk of Neonatal Death," *American Journal of Public Health*, 4(3), 411–414.

Bygren, L. O., Kaati, G., and Edvinsson, S.: (2001) "Longevity determined by ancestors' overnutrition during their slow growth period." *Acta Biotheoretica*, 49, 53–59.

Bygren, Olov (2010) in Kaati, G., Bygren, L. O., and Edvinsson, S.: (2002). "Cardiovascular and diabetes mortality determined by nutrition during parents' and grandparents' slow growth period." *European Journal of Human Genetics*, 10, 682–688.

Cairney, S. A., Lindsay, S., Paller, K. A., and Gaskell, M. G. (2017). "Sleep Preserves Original and Distorted Memory Traces." *Cortex*, 1–19.

Cairns, John, Overbaugh, Julie, and Miller, Stephan (1988). "The origin of mutants." *Nature*, 335, 142–145.

Callier, Viviane (2018). "Cells Talk and Help One Another via Tiny Tube Networks." *Quanta Magazine*.

Cannon, W. B. *The Wisdom of the Body*. (Washington: W. W. Norton, 1963).

Cantey, J. B., Bascik, S. L., Sánchez, P. J., et al. (2012). "Prevention of Mother-to-Infant Transmission of Influenza during the Postpartum Period." *American Journal of Perinatology*, 30(3), 233–240.

Cantin, M. and Genest, J. (1986). "The heart as an endocrine gland." *Clinical and Investigative Medicine*, 9(4), 319–327.

Cardamone, M. D., Tanasa, B., Perissi, V., et al. (2018). "Mitochondrial retrograde signaling in mammals is mediated by the transcriptional cofactor GPS2 via direct mitochondria-to-nucleus translocation." *Molecular Cell*, 69(5), 757–772.

Carpenter, S. (2012). "That gut feeling." *Monitor on Psychology* 43, 50–55.

Carson, B. S., Javedan, S. P., Freeman, J. M., Vining, E. P., Zuckerberg, A. L., Lauer, J. A., and Guarnieri, M. (1996). "Hemispherectomy: A hemidecortication approach and review of 52 cases." *Journal of Neurosurgery*, 84(6), 903–911.

Carvalho, R. N. and Gerdes, H. H. "Cellular Nanotubes: Membrane Channels for Intercellular Communication." In *Medicinal Chemistry and Pharmacological Potential of Fullerenes and Carbon Nanotubes* (363–372). (Dordrecht, Netherlands: Springer, 2008).

Caswell, J., Dotta, B., and Persinger, M. (2014). "Cerebral Biophoton Emission as a Potential Factor in Non-Local Human-Machine Interaction." *NeuroQuantology*, 12(1), 1–11.

Caudron, F. and Barral, Y. (2013). "A super-assembly of Whi3 encodes memory of deceptive encounters by single cells during yeast courtship." *Cell*, 155, 1244–1257.

CBS C-elegans (http://cbs.umn.edu/cgc/what-c-elegans).

CDC, Center for Disease Control and Prevention, Autism 2012. https://www.cdc.gov/ncbddd/autism/addm.html.

Center for Behavioral Health Statistics and Quality. *2014 National Survey on Drug Use and Health: Methodological Summary and Definitions*. Rockville, Md.: Substance Abuse and Mental Health Services Administration; 2015.

Chakravarthy, S. V. and Ghosh, J. (1997). "On Hebbian-like adaptation in heart muscle: A proposal for 'cardiac memory'." *Biological Cybernetics*, 76(3), 207–215.

Chalmers, David (1995). "Facing Up to the Problem of Consciousness." *Journal of Consciousness Studies*, 2(3), 200–219.

Chamberlain, David (1988). *Babies Remember Birth*. (Los Angeles: Jeremy P. Tarcher, 1988).

Chamberlain, David *Windows to the Womb*. (Berkeley, Calif.: North Atlantic Books, 2013).

Chan, W. F., Gurnot, C., Montine, T J., Nelson, J. L., et al. (2012). "Male microchimerism in the human female brain." *PLoS One*, 7(9), e45592. doi: 10.1371/journal.pone.0045592.

Charil, A., Laplante, D. P., Vaillancourt, C., and King, S. (2010). "Prenatal stress and brain development." *Brain Research Reviews*, 65, 56–79.

Chen, A. Y., Zhong, C., Lu, T. K. (2015). "Engineering living functional materials." *ACS Synthetic Biology* 4(1), 8–11.

Chen, Shanping, Cai, Diancai, Glanzman, David L., et al. (2014). "Reinstatement of long-term memory following erasure of its behavioral and synaptic expression in Aplysia." *eLife*, 3, e03896.

Chena, Yining, Mathesonb, Laura E., and Sakataa, Jon T. (2016). "Mechanisms underlying the social enhancement of vocal learning in songbirds." *Proceedings of the National Academy of Sciences*, published ahead of print May 31, 2016, doi:10.1073/pnas.1522306113. Thomas D. Albright (ed.), The Salk Institute for Biological Studies, La Jolla, Calif.

Chhatre, Sumedha, Metzger, David S., Nidich, Sanford, et al. (2013). "Effects of behavioral stress reduction Transcendental Meditation intervention in persons with HIV." *Journal of AIDS Care;* Psychological and Socio-medical Aspects of AIDS/HIV, 25(10), 1291–1297.

Chirinos, D. A., Ong, J. C., Garcini, L. M., Alvarado, D., and Fagundes, C. (2019). "Bereavement, self-reported sleep disturbances, and inflammation: Results from Project HEART." *Psychosomatic Medicine*, 81(1), 67–73.

Chong, Yolanda, Moffat, Jason, Andrews, Brenda J., et al. (2015). "Yeast Proteome Dynamics from Single Cell Imaging and Automated Analysis." *Cell,* 161, 1413–1424.

Chris Reid's web page: http://sydney.edu.au/science/biology/socialinsects/profiles/chris-reid.shtml.

Christianson, S. A. (1992). "Emotional stress and eyewitness memory: a critical review." *Psychological Bulletin*, 112(2), 284–309.

Chua, E. F. and Ahmed, R. (2016). "Electrical stimulation of the dorsolateral prefrontal cortex improves memory monitoring." *Neuropsychologia*, 85, 74–79.

Church, G. M., Gao, Y., and Kosuri, S. (2012). "Next-generation digital information storage in DNA." *Science*, 1226355.

Clarke, G., O'Mahony, S. M., Dinan, T. G., and Cryan, J. F. (2014). "Priming for health: Gut microbiota acquired in early life regulates physiology, brain and behavior." *Acta Paediatrica*, 103(8), 812–819.

Clayton, D. F. and London, S. E. (2014). "Advancing avian behavioral neuroendocrinology through genomics." *Frontiers in Neuroendocrinology*, 35(1), 58–71.

Clemens, L. E., Heldmaier, G., and Exner, C. (2009). "Keep cool: Memory is retained during hibernation in Alpine marmots." *Physiology & Behavior*, 98(1), 78–84.

Cohen, S. and Syme, S. (eds.). *Social Support and Health*. (Orlando, Fla.: Academic Press, 1985).

Cole, Steve W. (2009). "Social Regulation of Human Gene Expression." *Current Directions in Psychological Science*, 18(3), 132–137.

Coleman, I. (2005). "Low birth weight, developmental delays predict adult mental health disorders." Presentation at 3rd International Congress on Developmental Origins of Health and Disease.

Collier, Sandra. *Wake Up to Your Dreams*. (Toronto: Scholastic Canada, 1996)

Colombo, J. A. and Reisin, H. D. (2004) "Interlaminar astroglia of the cerebral cortex: a marker of the primate brain." *Brain Research* 1006, 126–131.

Conine, C. C., Sun, F., and Rando, O. J., et al. (2018). "Small RNAs gained during epididymal transit of sperm are essential for embryonic development in mice." *Developmental Cell,* 46(4), 470–480.

Connolly, J. D., Goodale, M. A., Menon, R. S., and Munoz, D. P. (2002). "Human fMRI evidence for the neural correlates of preparatory set." *Nature Neuroscience*, 5, 1345–1352.

Contera, S. *Nano Comes to Life: How Nanotechnology Is Transforming Medicine and the Future of Biology.* (Princeton, N.J.: Princeton University Press, 2019).

Conzen, Suzanne (2009). "Social isolation worsens cancer." *Institute for Genomics and Systems Biology.* Retrieved from http://www.igsb.org/news/igsb-fellow-suzanne-d.-conzen-social-isolation-in-mice-worsens-breast-cancer.

Cording, S., Hühn, J., Pabst, O., et al. (2013). "The intestinal micro-environment imprints stromal cells to promote efficient Treg induction in gut-draining lymph nodes." *Mucosal Immunology,* 7(2), 359–368.

Cornell, B. A., Braach-Maksvytis, V., King, L., Osman, P., Raguse, B., Wieczorek, L. and Pace, R. (1997). "A biosensor that uses ion-channel switches." *Nature*, 387, 580–583.

Corsi, D. J., Donelle, J., Sucha, E., et al. (2002). "Maternal cannabis use in pregnancy and child neurodevelopmental outcomes." *Nature Medicine.*

Costa, E., Chen, Y., Dong, E., Grayson, D. R., Kundakovic, M., and Guidotti, A. (2009). "GABAergic promoter hypermethylation as a model to study the neurochemistry of schizophrenia vulnerability." *Expert Review of Neurotherapeutics*, 9, 87–98. [PubMed: 19102671]

Craddock, T. J., Tuszynski, J. A., Priel, A., and Freedman, H. (2010). "Microtubule ionic conduction and its implications for higher cognitive functions." *Journal of Integrative Neuroscience*, 9(2), 103–122.

Craddock, T. J., Tuszynski, J. A., and Hameroff, S. (2012). "Cytoskeletal signaling: is memory encoded in microtubule lattices by CaMKII phosphorylation?" *PLoS Computational Biology.*

Crum, Alia J. and Langer, Ellen J. (2007). "Mind-Set Matters: Exercise and the Placebo Effect." *Psychological Science,* 18(2), 165–171.

Dadds, M.R., et al. (2006). "Attention to the eyes and fear-recognition deficits in child psychopathy." *British Journal of Psychiatry*, September, 189, 2801.

Dadds, Mark R.; Moul, Caroline; Dobson-Stone, Carol, et al. (2014). "Polymorphisms in the oxytocin receptor gene are associated with the development of psychopathy." *Development and Psychopathy,* 26(1), 21–31. doi: https://doi.org/10.1017/S0954579413000485.

Damasio, A. R. (1999). "The Feeling of What Happens: Body and Emotion in the Making of Consciousness." *New York Times Book Review,* 104, 8–8.

Darwin, Charles. *On the Origin of Species by Means of Natural Selection, or the Preservation of Favoured Races in the Struggle for Life.* Fellow of the Royal, Geological, Linnæan, etc. societies; Author of Journal of researches during HMS *Beagle*'s Voyage round the world. (London: John Murray, 1859).

Davidson, R. J., Kabat-Zinn, J., Sheridan, J. F., et al. (2003). "Alterations in brain and immune function produced by mindfulness meditation." *Psychosomatic Medicine,* 65(4), 564–570.

Dawson, S., Glasson, E. J., Dixon. G., and Bower, C. (2009). "Birth defects in children with autism spectrum disorders: A population-based, nested case-control study." *American Journal of Epidemiology*, 169(11), 1296–1303.

Day, N. L., Richardson, G. A., Geva, D., et al. (1994). "Effect of prenatal marijuana exposure on the cognitive development of children at age three." *Neurotoxicology and Teratology,* 16(2), 169–175.

Day, N., Sambamoorthi, U., Jasperse, D., et al. (1991). "Prenatal marijuana use and neonatal outcome." *Neurotoxicology and Teratology,* 13(3), 329–334.

De Craen, A. J., Roos, P. J., De Vries, A. L., and Kleijnen, J. (1996). "Effect of colour of drugs: Systematic review of perceived effect of drugs and of their effectiveness." *BMJ*, 313(7072), 1624–1626.

De Dreu, C. K. and Kret, M. E. (2016). "Oxytocin conditions intergroup relations through upregulated in-group empathy, cooperation, conformity, and defense." *Biological Psychiatry*, 79(3), 165–173.

de Kloet, A. D. and Herman, J. P. (2018). "Fat-brain connections: Adipocyte glucocorticoid control of stress and metabolism." *Frontiers in Neuroendocrinology*, 48, 50–57.

Delgado-García, J. M. (2015). "Cajal and the conceptual weakness of neural sciences." *Frontiers in Neuroanatomy*, 9, 128. doi: 10.3389/fnana.2015.00128.

Delgado-García, J. M. and Gruart, A. "Neural plasticity and regeneration: Myths and expectations," in *Brain Damage and Repair: From Molecular Research to Clinical Therapy*, T. Herdegen and J. M. Delgado-García, eds. (Dordrecht, Netherlands: Springer, 2004), 259–273.

DeMause, Lloyd (1982). "The Fetal Origins of History." *The Journal of Psychohistory*.

Demirtas-Tatlidede, Asli, Freitas, Catarina, Cromer, Jennifer R., et al. (2010). "Safety and Proof of Principle Study of Cerebellar Vermal Theta Burst Stimulation in Refractory Schizophrenia." *Schizophrenia Research*, 2010 124(1–3), 91–100, published online August 22, 2014. doi: 10.1093/brain/awu239 PMCID: PMC461413.

Denk, F., Crow, M., Didangelos, A., Lopes, D. M., and McMahon, S. B. (2016). "Persistent alterations in microglial enhancers in a model of chronic pain." *Cell Reports*, 15(8), 1771–1781.

Dennett, D. C. *Consciousness Explained*. (Boston: Little, Brown, 1991).

Derecki, N. C., Cardani, A. N., Yang, C. H., Quinnies, K. M., Crihfield, A., Lynch, K. R., and Kipnis, J. (2010). "Regulation of learning and memory by meningeal immunity: A key role for IL-4." *Journal of Experimental Medicine*, 207(5), 1067–1080.

Dermitzakis, Emmanouil T., et al. (2015). "Population Variation and Genetic Control of Modular Chromatin Architecture in Humans." *Cell*, 162(5), 1039–1050.

Diamanti-Kandarakis, E., Bourguignon, J. P., Giudice, L. C., Hauser, R., Prins, G. S., Soto, A. M., and Gore, A. C. (2009). "Endocrine-disrupting chemicals: An Endocrine Society scientific statement." *Endocrine Reviews*, 30(4), 293–342.

Dias, B. G., Maddox, S. A., Klengel, T., and Ressler, K. J. (2015). "Epigenetic mechanisms underlying learning and the inheritance of learned behaviors." *Trends in Neurosciences*, 38(2), 96–107.

Dickerson, F., Adamos, M., and Yolken, R. H. (2018)." Adjunctive probiotic microorganisms to prevent rehospitalization in patients with acute mania: A randomized controlled trial." *Bipolar Disorders*, 20(7), 614–621.

Dickson, D. A., Paulus, J. K., Feig, L. A., et al. (2018). "Reduced levels of miRNAs 449 and 34 in sperm of mice and men exposed to early life stress." *Translational Psychiatry*, 8(1), 1–10.

Dietz, D. M., LaPlant, Q., Watts, E. L., Hodes, G. E., Russo, S. J., Feng, J., et al. (2011): "Paternal transmission of stress-induced pathologies." *Biological Psychiatry*, 70, 408–414.

Dietz, David M. and Nestler, Eric J. (2012). "From Father to Offspring: Paternal Transmission of Depressive-Like Behaviors." *Neuropsychopharmacology Reviews*, 37, 311–312. doi:10.1038/npp.2011.16.

Dinan, T. G. and Cryan, J. F. (2013). "Melancholic microbes: A link between gut microbiota and depression?" *Neurogastroenterology and Motility*, 25(9), 713–719.

Dobbing, J. and Sands, J. (1973). "Quantitative growth and development of human brain." *Archives of Diseases of Childhood*, 48, 757–767.

Dodonova, S. O., Diestelkoetter-Bachert, P., Beck, R., Beck, M., et al. (2015). "A structure of the COPI coat and the role of coat proteins in membrane vesicle assembly." *Science*, 349(6244), 195.

Donkin, I. and Barrès, R. (2018). "Sperm epigenetics and influence of environmental factors." *Molecular Metabolism*, 14, 1–11.

Dörner, T. and Radbruch, A. (2007). "Antibodies and B cell memory in viral immunity." *Immunity*, 27(3), 384–392.

Dubey, G. P. and Ben-Yehuda, S. (2011). "Intercellular nanotubes mediate bacterial communication." *Cell*, 144(4), 590–600.

Dudas, M., Wysocki, A., Gelpi, B. and Tuan, T. L. (2008). "Memory encoded throughout our bodies: Molecular and cellular basis of tissue regeneration." *Pediatric Research*, 63(5), 502–512.

Duffy, J. F., Cain, S. W., Chang, A. M., Phillips, A. J., Münch, M. Y., Gronfier, C., and Czeisler, C. A. (2011). "Sex difference in the near-24-hour intrinsic period of the human circadian timing system." *Proceedings of the National Academy of Sciences*, 108 (Supplement 3), 15602–15608.

Dworkin, M. (1983). "Tactic behavior of Myxococcus xanthus." *Journal of Bacteriology*, 154, 452–459.

Ebert, M. S. and Sharp, P. A. (2012) "Roles for microRNAs in conferring robustness to biological processes." *Cell*, 149, 515–524.

Eccles, J. C. (1992). "Evolution of consciousness." *Proceedings of the National Academy of Sciences*, 89, 7320–7324.

Egner, I. M., Bruusgaard, J. C., Eftestøl, E., and Gundersen, K. (2013). "A cellular memory mechanism aids overload hypertrophy in muscle long after an episodic exposure to anabolic steroids." *The Journal of Physiology*, 591(24), 6221–6230.

Ehlen, J. C., Brager, A. J., Baggs, J., Pinckney, L., Gray, C. L., DeBruyne, J. P., Paul, K. N., Esser, K. A., Joseph, S., and Takahashi, J. S. (2017). "Bmal1 function in skeletal muscle regulates sleep." *eLife*, 6.

El Marroun, H., Tiemeier, H., Steegers, E. A., and Huizink, A. C. (2009). "Intrauterine cannabis exposure affects fetal growth trajectories: the Generation R Study." *Journal of the American Academy of Child & Adolescent Psychiatry*, 48(12), 1173–1181.

El-Athman, R., Genov, N. N., Mazuch J., Zhang, K., Yu, Y., Fuhr, L., Relógio, A., et al. (2017). "The Ink4a/Arf locus operates as a regulator of the circadian clock modulating RAS activity." *PLoS Biology*, 15(2), e2002940.

Elsevier (2014). "Discovery of quantum vibrations in 'microtubules' corroborates theory of consciousness." Retrieved from https://phys.org/news/2014-01-discovery-quantum-vibrations-microtubules-corroborates.html#jCp.

Emoto, Masaru. *The Hidden Message in Water.* (New York: Atria Books, 2005).

Epel, E., Blackburn, E., Lin, J., et al. (2004). "Accelerated telomere shortening in response to exposure to life stress." *Proceedings of the National Academy of Sciences*, 101, 17312–17315. [PubMed: 15574496]

Epel, E. S., Lin, J., Wilhelm, F. H., et al. (2006). "Cell aging in relation to stress arousal and cardiovascular disease risk factors." *Psychoneuroendocrinology*, 31(3), 277–287. [PubMed: 16298085]

Epel, E., Daubenmier, J., Moskowitz, J. T., Folkman, S., and Blackburn, E. (2009). "Can meditation slow rate of cellular aging? Cognitive stress, mindfulness, and telomeres." *Annals of the New York Academy of Sciences*, 1172(1), 34–53.

Essex, Marilyn J., Boyce, W., Thomas, Kobor, and Michael, S. (2011). "Epigenetic Vestiges of Early Developmental Adversity: Childhood Stress Exposure and DNA Methylation in Adolescence." *Child Development*. doi: 10.1111/j.1467-8624.2011.01641.

European College of Neuropsychopharmacology. "Scientists find psychiatric drugs affect gut contents." *ScienceDaily*, September 9, 2019.

European Society of Cardiology (June 9, 2018). "Loneliness is bad for the heart." *ScienceDaily*. Retrieved June 10, 2018.

Evertz, Klaus, Janus, Ludwig, and Linder, Rupert. *Lehrbuch der Prenatalen Psychologie*. (Heidelberg, Germany: Mattes Verlag, 2014).

Evolution News & Science Today (EN) (2018). "Memory—New Research Reveals Cells Have It, Too." https://evolutionnews.org/2018/11/memory-new-research-reveals-cells-have-it-too/.

Extremophile: In *Wikipedia*: https://en.wikipedia.org/wiki/Extremophile.

Eyal, G., Verhoog, M. B., Testa-Silva, G., Deitcher, Y., Lodder, J. C., Benavides-Piccione, R., and Segev, I. (2016). "Unique membrane properties and enhanced signal processing in human neocortical neurons." *eLife*, 5, e16553.

Faa, G., Gerosa, C., Fanni, D., Nemolato, S., van Eyken, P., and Fanos, V. (2013). "Factors influencing the development of a personal tailored microbiota in the neonate, with particular emphasis on antibiotic therapy." *The Journal of Maternal-Fetal & Neonatal Medicine*, 26(sup2), 35–43.

Fahrbach, S. E. (2006). "Structure of the mushroom bodies of the insect brain." *Annual Review of Entomology*, 51, 209–32.

Fang, C. Y., Reibel, D. K., Longacre, M. L., Douglas, S. D., et al. (2010). "Enhanced psychosocial well-being following participation in a mindfulness-based stress reduction program is associated with increased natural killer cell activity." *The Journal of Alternative and Complementary Medicine*, 16(5), 531–538.

Faraco, G., Brea, D., Sugiyama, Y., et al. (2018). "Dietary salt promotes neurovascular and cognitive dysfunction through a gut-initiated TH17 response." *Nature Neuroscience*, 21(2), 240–249.

Farrant, Graham (1988). "An Interview with Dr. Graham Farrant by Steven Raymond." *Pre- & Perinatal Psychology News*, 2(2) (Summer).

Fauci, Anthony S. and Harrison, T. R., eds. *Harrison's Principles of Internal Medicine*. (17th ed.). (New York: McGraw-Hill Medical, 2008), 2339–2346.

Feng Yu, Qing-jun Jiang, Xi-yan Sun, and Rong-wei Zhang (2015). "A new case of complete primary cerebellar agenesis: Clinical and imaging findings in a living patient." *Brain*, 138(6), e353. doi: 10.1093/brain/awu239.

Fessenden, Jim (2018). "Changes to small RNA in sperm may help fertilization. Research in mouse embryos sheds new light on inheritance of traits." *UMass Medical School Communications*, https://www.umassmed.edu/news/news-archives/2018/07/changes-to-small-rna-in-sperm-may-help-fertilization/.

Feuillet, L., Dufour, H., and Pelletier, J. (2007). "Brain of a white-collar worker." *Lancet*, 370(9583), 262.

Fields, R. Douglas (2010). "Visualizing Calcium Signaling in Astrocytes." *Science Signaling*, 3(147).

Fields, R. Douglas. *The Other Brain*. (New York: Simon & Schuster, 2011).

Fields, R. Douglas (2013). "Human Brain Cells Make Mice Smart." *Scientific American*.

Fields, Douglas (2013). "'Brainy' Mice with Human Brain Cells: Chimeras of Mice and Men." *BrainFacts/Society for Neuroscience*.

Filiano, Anthony J., Litvak, Vladimir, Kipnis, Jonathan, et al. (2016). "Unexpected role of interferon-γ in regulating neuronal connectivity and social behavior." *Nature*, 535(7612), 425–429.

Filiano, A. J., Gadani, S. P., and Kipnis, J. (2015). "Interactions of innate and adaptive immunity in brain development and function." *Brain Research*, 1617, 18–27.

Finigan, V. (2012). "Breastfeeding and diabetes: Part 2." *The Practising Midwife*, 15(11), 33–34, 36.

Flanagan, Owen. *The Science of the Mind*. 2nd ed. (Cambridge, Mass.: MIT Press, 1991).

Flinders University (2020). Biofilm Research and Innovation Consortium. https://www.flinders.edu.au/biofilm-research-innovation-consortium.

Fogel, Alan (2010). "PTSD is a chronic impairment of the body sense: Why we need embodied approaches to treat trauma in Body Sense," online blog from *Psychology Today* https://www.psychologytoday.com/blog/body-sense/201006/ptsd-is-chronic-impairment-the-body-sense-why-we-need-embodied-approaches.

Fogelman, E. and Savran, B. (1980). "Brief group therapy with offspring of Holocaust survivors: Leaders' reactions." *American Journal of Orthopsychiatry*, 50(1), 96.

Forssmann, W. G., Hock, D., Lottspeich, F., Henschen, A., Kreye, V., Christmann, M., and Mutt, V. (1983). "The right auricle of the heart is an endocrine organ." *Anatomy and Embryology*, 168(3), 307–313.

Foster, P. L. (1999). "Mechanisms of stationary phase mutation: A decade of adaptive mutation." *Annual Review of Genetics*, 33, 57–88.

Frank, Adam (2017). "Minding matter." *Aeon Magazine.*

Franklin, Robin J. M. and Bussey, Timothy J. (2013). "Do Your Glial Cells Make You Clever?" *Cell Stem Cell.*

Franklin, Tamara B., Saab, Bechara J., and Mansuy, Isabelle M. (2012). "Neural Mechanisms of Stress Resilience and Vulnerability." *Neuron*, 75(5), 747–761.

Fredrickson, B. L., Grewen, K. M., Cole, S. W., et al. (2013). "A functional genomic perspective on human well-being." *Proceedings of the National Academy of Sciences*, 110(33), 13684–13689.

Freitas, Joao (2017). "Teacher Shows Students How Negative Words Can Make Rice Moldy." Retrieved from https://www.goodnewsnetwork.org/teacher-shows-students-how-negative-words-makes-rice-moldy/.

Freud, A. and Burlingham, D. *War and Children.* (New York: International Universities Press, 1942).

Freyberg, J. T. (1980). "Difficulties in separation-individuation as experienced by holocaust survivors." *American Journal of Orthopsychiatry*, 50(1), 87–95.

Fried, P. A. "Pregnancy." In F. Grotenhermen and E. Russo (eds.), *Cannabis and cannabinoids: Pharmacology, toxicology, and therapeutic potential* (New York: Haworth Integrative Healing Press, 2002), 269–278.

Fried, P. A. and Watkinson, B. (1990). "36- and 48-month neurobehavioral follow-up of children prenatally exposed to marijuana, cigarettes, and alcohol." *Journal of Developmental and Behavioral Pediatrics*, 11(2), 49–58.

Fuchs, T. "Body memory and the unconscious." In *Founding Psychoanalysis Phenomenologically* (Dordrecht, Netherlands: Springer, 2012), 69–82.

Gaidos, Susan (2013). "Memories lost and found: Drugs that help mice remember reveal role for epigenetics in recall." *Science News.*

Gallwey, W. Timothy. *The Inner Game of Tennis: The Classic Guide to the Mental Side of Peak.* (New York: Random House, 1997).

Gandhi, S. R., Yurtsev, E. A., Korolev, K. S., and Gore, J. (2016). "Range expansions transition from pulled to pushed waves as growth becomes more cooperative in an experimental microbial population." *Proceedings of the National Academy of Sciences*, 113(25), 6922–6927.

Gapp, Katharina, Bohacek, Johannes, Mansuy, Isabelle M., et al. (2016). "Potential of Environmental Enrichment to Prevent Transgenerational Effects of Paternal Trauma." *Neuropsychopharmacology.* doi: 10.1038/npp.2016.87.

Gapp, Katharina, Jawaid, Ali, Sarkies, Peter, Mansuy, Isabelle M., et al. (2014). "Implication of sperm RNAs in transgenerational inheritance of the effects of early trauma in mice." *Nature Neuroscience*, 17, 667–669; doi:10.1038/nn.3695.

Gary Marcus in Cepelewicz, Jordana (January 14, 2020). "Hidden Computational Power Found in the Arms of Neurons." *Quanta Magazine.* Retrieved from https://www.quantamagazine.org/neural-dendrites-reveal-their-computational-power-20200114/.

Gattinoni, L., Lugli, E., Restifo, N. P. (2011). "A human memory T cell subset with stem cell-like properties." *Nature Medicine*, 18;17(10), 1290–1297.

Gaydos, Laura J., Wang, Wenchao, and Strome, Susan (2014). "H3K27me and PRC2 transmit a memory of repression across generations and during development. *Science*, 345(6203), 1515–1518. doi: 10.1126/science.1255023.

Gazzaniga, Michæl S. *The Mind's Past.* (Berkeley: University of California Press, 2000).

Gebauer, Juliane, Gentsch, Christoph, Kaleta, Christoph, et al. (2016). "A Genome-Scale Database and Reconstruction of Caenorhabditis elegans Metabolism." *Cell Systems*, 2(5), 312. doi:10.1016/j.cels.2016.04.017.

Gentile, L., Cebria, F.. and Bartscherer, K. (2011) "The planarian flatworm: An in vivo model for stem cell biology and nervous system regeneration." *Disease Models & Mechanisms*, 4(1), 12-9.

Gershon, M. D. (2020). "The Thoughtful Bowel." *Acta Physiologica*, 228(1), e13331.

Gershon, Michael D., *The Second Brain.* (New York: HarperCollins, 1998).

Gidon, A., Zolnik, T. A., Larkum, M. E., et al. (2020). "Dendritic action potentials and computation in human layer 2/3 cortical neurons." *Science*, 367(6473), 83–87.

Gilbert, S. F. (2003). "Developmental Biology." Sinauer Associates, Inc, Sunderland, 1–838.

Glassman, A. H., Bigger, J. T., and Gaffney, M. (2007). "Heart Rate Variability in Acute Coronary Syndrome Patients with Major Depression, influence of Sertraline and Mood Improvement." *Archives of General Psychiatry*, 64, 9.

Glennie, N. D., Yeramilli, V. A., Beiting, D. P., Volk, S. W., Weaver, C. T., and Scott, P. (2015). "Skin-resident memory CD4+ T cells enhance protection against Leishmania major infection." *Journal of Experimental Medicine*, jem-20142101.

Goldilocks and the Three Bears. [June 17, 2017]. In *Wikipedia*. Retrieved June 21, 2017, from https://en.wikipedia.org/w/index.php?title=Goldilocks_and_the_Three_Bears&oldid=786195942).

Golding, Jean, Ellis, Genette, Pembrey, Marcus, et al. (2017). "Grand-maternal smoking in pregnancy and grandchild's autistic traits and diagnosed autism." *Scientific Reports* 7, article number 46179.

Goldman, N., Bertone, P., Chen, S., Dessimoz, C., LeProust, E. M., Sipos, B., and Birney, E. (2013). "Towards practical, high-capacity, low-maintenance information storage in synthesized DNA." *Nature*, 494(7435), 77–80.

Goldstein, P., Weissman-Fogel, I., Dumas, G., and Shamay-Tsoory, S. G. (2018). "Brain-to-brain coupling during handholding is associated with pain reduction." *Proceedings of the National Academy of Sciences*, 201703643.

Gonzalez, Walter G., Zhang, Hanwen, Harutyunyan, Anna, Lois, Carlos (2019). "Persistence of neuronal representations through time and damage in the hippocampus." *Science*, 365(6455), 821–825.

Gordon, I., Gilboa, A., Siegman, S., et al. (2020). "Physiological and Behavioral Synchrony predict Group cohesion and performance." *Scientific Reports*, 10(1), 1–12.

Grace, Katherine (2017). "30 days of love, hate and indifference: Rice and water experiment #1." Retrieved from https://yayyayskitchen.com/2017/02/02/30-days-of-love-hate-and-indifference-rice-and-water-experiment-1/.

Gräff, J, and Tsai, L. H. (2013). "Histone acetylation: molecular mnemonics on the chromatin." *Nature Reviews Neuroscience*, 14, 97–111.

Greenspan, R. J. *An Introduction to Nervous Systems*. (Cold Spring Harbor, N.Y.: Cold Spring Harbor Laboratory Press, 2007).

Grémiaux, A., Yokawa, K., Mancuso, S., and Baluška, F. (2014). "Plant anesthesia supports similarities between animals and plants." *Claude Bernard's Forgotten Studies*, 9, e27886.

Grenham, S., Clarke, G., Cryan, J. F., and Dinan, T. G. (2011). "Brain–gut–microbe communication in health and disease." *Frontiers in Physiology*, *2*.

Gribble, Karleen D. (2006). "Mental health, attachment and breastfeeding: Implications for adopted children and their mothers." *International Breastfeeding Journal*, 1, 5.

Griesler, P. C., Kandel, D. B., and Davies, M. (1998). "Maternal smoking in pregnancy, child behavior problems, and adolescent smoking." *Journal of Research on Adolescence*; 8, 159–185. doi:10.1207/s15327795jra0802_1.

Griffin, Matthew (August 29, 2016). "Researchers Find Evidence That Ancestors Memories Are Passed Down in DNA." *Enhanced Humans and Biotech*. https://www.311institute.com/researchers-find-evidence-that-ancestors-memories-are-passed-down-in-dna/.

Grossmann, Klaus E., Grossmann, Karin, and Waters, Everett. *Attachment from Infancy to Adulthood: The Major Longitudinal Studies.* (New York: Guilford Press, 2005).

Gu, Y., Vorburger, R. S., Gazes, Y., Habeck, C. G., Stern, Y., Luchsinger, J. A., and Brickman, A. M. (2016). "White matter integrity as a mediator in the relationship between dietary nutrients and cognition in the elderly." *Annals of Neurology*, 79(6), 1014–1025.

Guan, Q., Haroon, S., Bravo, D. G., Will, J. L., and Gasch, A. P. (2012). "Cellular memory of acquired stress resistance in Saccharomyces cerevisiae." *Genetics*, 192(2), 495–505.

Gulliver's Travels. (n.d.) In *Wikipedia.* Retrieved August 2, 2017, from https://en.wikipedia.org/wiki/Gulliver%27s_Travels.

Hackett, Jamie A., Sengupta, Roopsha, Surani, M. Azim, et al. (2013). "Germline DNA demethylation dynamics and imprint erasure through 5-hydroxymethylcytosine." *Science*, 339(6118), 448–452, American Association for the Advancement of Science.

Hadj-Moussa, H. and Storey, K. B. (2018). "Micromanaging freeze tolerance: The biogenesis and regulation of neuroprotective microRNAs in frozen brains." *Cellular and Molecular Life Sciences*, 75(19), 3635–3647.

Hameroff, S. and Chopra, D. "The 'quantum soul': A scientific hypothesis." In Moreira-Almeida, A. and Santos, F. S. (eds.), *Exploring Frontiers of the Mind–Brain Relationship.* (New York: Springer, 2012).

Hameroff, Stuart (2014). "Consciousness, Microtubules, & 'Orch OR' : A 'Space-time Odyssey.'" *Journal of Consciousness Studies*, 21(3–4), 126–158.

Hameroff, Stuart (November 2015). "Is your brain really a computer, or is it a quantum orchestra tuned to the universe?" *Interalia Magazine.* Retrieved from https://www.interaliamag.org/articles/stuart-hameroff-is-your-brain-really-a-computer-or-is-it-a-quantum-orchestra-tuned-to-the-universe/.

Hamilton, T. C. "Behavioral plasticity in protozoans," in *Aneural Organisms in Neurobiology*, ed. Eisenstein, E. M. (New York: Plenum Press, 1975), 111–130.

Han, X., Chen, M., Wang, F., Windrem, M., Wang, S., Shanz, S., and Silva, A. J. (2013). "Forebrain engraftment by human glial progenitor cells enhances synaptic plasticity and learning in adult mice." *Cell Stem Cell*, 12(3), 342–353.

Han, W., Tellez, L. A., Kaelberer, M. M., et al. (2018). "A neural circuit for gut-induced reward." *Cell*, 175(3), 665–678.

Hartsough, L. A., Kotlajich, M. V., Tabor, J. J., et al. (2020). "Optogenetic control of gut bacterial metabolism." *bioRxiv.*

Harvard Public Health Magazine (2011). Retrieved from hsph.harvard.edu/news/magazine/happiness-stress-heart-disease/ (2011).

Hauck, F. R., Thompson, J. M., Tanabe, K. O., Moon, R. Y., Vennemann, M. M. (2011). "Breastfeeding and reduced risk of sudden infant death syndrome: A meta-analysis." *Pediatrics*, 128(1), 103–10.

Hazelbauer, G. L., Falke, J. J., and Parkinson, J. S. (2007). "Bacterial chemoreceptors: High-performance signaling in networked arrays." *Trends in Biochemical Sciences*, 33, 9–19.

Heather Ross in De Giorgio Lorriana (March 28, 2012). "Can a heart transplant change your personality?" *Toronto Star.* Retrieved from https://www.thestar.com/news/world/2012/03/28.

Heim, C. and Nemeroff, C. B. (2001). "The role of childhood trauma in the neurobiology of mood and anxiety disorders: preclinical and clinical studies." *Biological Psychiatry*, 49, 1023–1039. [PubMed: 11430844]

Heler, R., Samai, P., Modell, J. W., Weiner, C., Goldberg, G. W., Bikard, D., and Marraffini, L. A. (2015). "Cas9 specifies functional viral targets during CRISPR-Cas adaptation." *Nature*, 519(7542), 199.

Hennekens, C. H., Hennekens, A. R., Hollar, D., and Casey, D. E. (2005). "Schizophrenia and increased risks of cardiovascular disease." *American Heart Journal*, 150(6), 1115–1121.

Hepper, P. G. and Waldman, B. (1992). "Embryonic olfactory learning in frogs." *Quarterly Journal of Experimental Psychology*, Section B, 44(3–4), 179–197.

Hergenhahn, B. and Henley, T. *An Introduction to the History of Psychology*. (Andover, UK: Cengage Learning, 2013).

Herman, James (September 12, 2019). In "Fight or Flight May Be in Our Bones," by Diana Kwon. *Scientific American*.

Heron, J., O'Connor, T. G., Evans, J., Golding, J., Glover, V., and ALSPAC Study Team. (2004). "The course of anxiety and depression through pregnancy and the postpartum in a community sample." *Journal of Affective Disorders*, 80(1), 65–73.

Hezroni, Hadas, Perry, Rotem Ben-Tov, Ulitsky, Igor, et al. (2017). "A subset of conserved mammalian long non-coding RNAs are fossils of ancestral protein-coding genes." *Genome Biology*, 18(1).

Hibberd, T. J., Yew, W. P., Chen, B. N., Costa, M., Brookes, S. J., and Spencer, N. J. (2020). "A novel mode of sympathetic reflex activation mediated by the enteric nervous system." *eNeuro*.

"Hippo Tsunami Survivor: A True Story of Survival." (2005). Retrieved from https://inspire21.com/hippo-tsunami-survivor/.

Hittner, E. F., Stephens, J. E., Turiano, N. A., Gerstorf, D., Lachman, M. E., and Haase, C. M. (2020). "Positive Affect Is Associated with Less Memory Decline: Evidence From a 9-Year Longitudinal Study." *Psychological Science,* 0956797620953883.

Ho, B., Baryshnikova, A., and Brown, G. W. (2018). "Unification of protein abundance datasets yields a quantitative Saccharomyces cerevisiae proteome." *Cell Systems*, 6(2), 192–205.

Hoff, E. (2003). "The specificity of environmental influence: Socioeconomic status affects early vocabulary development via maternal speech." *Child Development,* 2;74, 1368–1878.

Hoff, E. (2006); "How social contexts support and shape language development." *Developmental Review*, 26, 55–88.

Holleran, Laurena, Kelly, Sinead, Donoho, Gary, et al. (2020). "The Relationship Between White Matter Microstructure and General Cognitive Ability in Patients With Schizophrenia and Healthy Participants in the ENIGMA Consortium." *American Journal of Psychiatry*, appi.ajp.2019.1.

Holden, C. (2005). "Sex and the suffering brain." *Science*, 308, 1574–1577.

Holdevici, I. (2014). "A brief introduction to the history and clinical use of hypnosis." *Romanian Journal of Cognitive Behavioral Therapy and Hypnosis*, 1(1), 1–5.

Hu, S., Dong, T. S., Dalal, S. R., Wu, F., Bissonnette, M., Kwon, J. H., and Chang, E. B. (2011). "The microbe-derived short chain fatty acid butyrate targets miRNA-dependent p21 gene expression in human colon cancer." *PloS One*, 6(1), e16221.

Hudetz, A. and Pearce, R. (eds.) (2010). *Suppressing the Mind: Anesthetic Modulation of Memory and Consciousness*. (Totowa, N.J.: Humana Press, 2010).

Humphreys, Lloyd G. (1979). "The construct of general intelligence." *Intelligence*, 3(2), 105–120.

Humphries, M. M., Thomas, D. W., and Kramer, D. L. (2003). "The role of energy availability in mammalian hibernation: A cost-benefit approach." *Physiological and Biochemical Zoology*, 76(2), 165–179.

Hus, S. M., Ge, R., Akinwande, D., et al. (2020). "Observation of single-defect memristor in an MoS 2 atomic sheet." *Nature Nanotechnology*, 1–5.

Ignaszewski, M. J., Yip, A., and Fitzpatrick, S. (2015). "Schizophrenia and coronary artery disease." *British Columbia Medical Journal*, 57(4), 154–157.

Ingalhalikar, M., Smith, A., Parker, D., Satterthwaite, T. D., Elliott, M. A., Ruparel, K., and Verma, R. (2014). "Sex differences in the structural connectome of the human brain." *Proceedings of the National Academy of Sciences*, 111(2), 823–828.

Inoue, J. (2008). "A simple Hopfield-like cellular network model of plant intelligence." *Progress in Brain Research*, 168, 169–174.

Inoue, T., Kumamoto, H., Okamoto, K., Umesono, Y., Sakai, M., Sánchez Alvarado, A., and Agata, K. (2004). "Morphological and functional recovery of the planarian photosensing system during head regeneration." *Zoological Science*, 21, 275–283.

Inspector, Y., Kutz, I., and Daniel, D. (2004). "Another person's heart: magical and rational thinking in the psychological adaptation to heart transplantation." *The Israel Journal of Psychiatry and Related Sciences*, 41(3), 161.

Islam, M. R., Lbik, D., Sakib, M. S., Maximilian Hofmann, R., Berulava, T., Jiménez Mausbach, M., and Fischer, A. (2021). "Epigenetic gene expression links heart failure to memory impairment." *EMBO Molecular Medicine*, 13(3), e11900.

Jab, Ferris (2012). "Know Your Neurons: What Is the Ratio of Glia to Neurons in the Brain?" *Scientific American*.

Jacka, Felice N., Ystrom, Eivind, Brantsaeter, Anne Lise, et al. (2013). "Maternal and Early Postnatal Nutrition and Mental Health of Offspring by Age 5 Years: A Prospective Cohort Study." *Journal of the American Academy of Child & Adolescent Psychiatry*.

Jadhav, U., Cavazza, A., Banerjee, K. K., Xie, H., O'Neill, N. K., Saenz-Vash, V., and Shivdasani, R. A. (2019). "Extensive recovery of embryonic enhancer and gene memory stored in hypomethylated enhancer DNA." *Molecular Cell*, 74(3), 542–554.

James Shapiro in Ball, P. (2008). "Cellular memory hints at the origins of intelligence." *Nature*, 451, 385.

James, William (1884). "What Is an Emotion?" *Mind*, os-IX(34), 188–205.

James, William. "What Is an Emotion" in Richardson, R. D. (ed.). *The Heart of William James*. (Boston: Belknap Press, 2012).

Janik, R., Thomason, L. A., Stanisz, A. M., Forsythe, P., Bienenstock, J., and Stanisz, G. J. (2016). "Magnetic resonance spectroscopy reveals oral Lactobacillus promotion of increases in brain GABA, N-acetyl aspartate and glutamate." *Neuroimage*, 125, 988–995.

Jensen, M. P., Jamieson, G. A., Santarcangelo, E. L., Terhune, D. B., et al. (2017). "New directions in hypnosis research: strategies for advancing the cognitive and clinical neuroscience of hypnosis." *Neuroscience of Consciousness*, 2017(1), nix004.

Jindal, R., MacKenzie, E. M., Baker, G. B., and Yeragani, V. K. (2005). "Cardiac risk and schizophrenia." *Journal of Psychiatry and Neuroscience*, 30(6), 393.

Jobson, M. A., Jordan, J. M.; Sandrof, M. A., Baugh. L. R., et al. (2015). "Transgenerational Effects of Early Life Starvation on Growth, Reproduction and Stress Resistance in Caenorhabditis elegans." *Genetics*. doi: 10.1534/genetics.115.178699.

John Schroeder in Skeptic's Dictionary Online: http://skepdic.com/cellular.html.

Johnson, George (2016). "Physicists Recover From a Summer's Particle 'Hangover'." *New York Times*.

Jose, A. M. (2020). "Heritable Epigenetic Changes Alter Transgenerational Waveforms Maintained by Cycling Stores of Information." *BioEssays*, 1900254.

Jose, A. M. (2020). "A framework for parsing heritable information." *Journal of the Royal Society Interface*, 17(165), 20200154.

June, Catharine (2015). "Michigan Micro Mote (M3) makes history as the world's smallest computer." https://ece.engin.umich.edu/stories/michigan-micro-mote-m3-makes-history-as-the-worlds-smallest-computer.

Jung, C. G. *Archetypes and the Collective Unconscious (The Collected works of C.G. Jung Vol 1, Pt 1)*. (Princeton, N.J.: Princeton University Press, 1969).

Kacsoh, B. Z., Bozler, J., Ramaswami, M., and Bosco, G. (2015). "Social communication of predator-induced changes in Drosophila behavior and germ line physiology." *eLife*, 4, e07423.

Kaliman, Perla, Alvarez-Lopez, Maria Jesus, Davidson, Richard J., et al. (2014). "Rapid changes in histone deacetylases and inflammatory gene expression in expert meditators." *Psychoneuroendocrinology,* 40, 96–107.

Kandel, Eric R. (2002). "The Molecular Biology of Memory Storage: A Dialog Between Genes and Synapses." *Bioscience Reports,* 21(5). Plenum Publishing Corporation, 567.

Kanduri, Chakravarthi, Raijas, Pirre, and Järvelä, Irma (2015). "The effect of listening to music on human transcriptome." *PeerJ,* 23, e830. doi: 10.7717/peerj.830.

Kauffman, S. (2010). "Is There A 'Poised Realm' Between the Quantum and Classical Worlds?" *Cosmos and Culture.*

Kauffman, S. *Humanity in a Creative Universe.* (Oxford, UK: Oxford University Press, 2016).

Kauffman, S., Niiranen, S., and Vattay, G. (2014). *U.S. Patent No. 8,849,580.* Washington, DC: U.S. Patent and Trademark Office.

Kaufman J., Plotsky, P. M., Nemeroff, C. B., Charney, D. S. (2000). "Effects of early adverse experiences on brain structure and function: Clinical implications." *Biological Psychiatry,* 48, 778–790. [PubMed:11063974]

Kirkey, Sharon. (2019). "The rise of 'psychobiotics?' 'Poop pills' and probiotics could be game changers for mental illness." *National Post,* October 8, 2019.

Kern, Elizabeth M.A., Robinson, Detric, Langerhans, R. Brian, et al. (2016). "Correlated evolution of personality, morphology and performance." *Animal Behavior,* 117, 79. doi: 10.1016/j.anbehav.2016.04.007.

Khan, S. M., Ali, R., Asi, N., and Molloy, J. E. (2012). "Active actin gels." *Communicative & Integrative Biology.*

Kikkert, S., Kolasinski, J., Jbabdi, S., Tracey, I., Beckmann, C. F., Johansen-Berg, H., and Makin, T. R. (2016). "Revealing the neural fingerprints of a missing hand." *eLife,* 5, e15292.

Knowles, J. R. (1980). "Enzyme-catalyzed phosphoryl transfer reactions." *Annual Review of Biochemistry,* 49(1), 877–919.

Kim, M. K., Ingremeau, F., Zhao, A., Bassler, B. L., and Stone, H. A. (2016). "Local and global consequences of flow on bacterial quorum sensing." *Nature Microbiology,* 1(1), 1–5.

Koenig, J. E., Spor, A., Scalfone, N., Fricker, A. D., Stombaugh, J., Knight, R., and Ley, R. E. (2011). "Succession of microbial consortia in the developing infant gut microbiome." *Proceedings of the National Academy of Sciences,* 108(supplement 1), 4578–4585.

Korosi A. and Baram, T. Z. (2009). "The pathways from mother's love to baby's future." *Frontiers in Behavioral Neuroscience.* Epub ahead of print September 24, 2009.

Koshland Jr., D. E. (1980). "Bacterial chemotaxis in relation to neurobiology." *Annual Review of Neuroscience,* 3(1), 43–75; and Lyon, P. (2015). "The cognitive cell: bacterial behavior reconsidered." *Frontiers in Microbiology,* 6.

Kubzansky, L. D. and Thurston, R. C. (2007). "Emotional vitality and incident coronary heart disease: Benefits of healthy psychological functioning." *Archives of General Psychiatry,* 64(12), 1393–1401.

Kugathasan, P., Johansen, M. B., Jensen, M. B., Aagaard, J., and Jensen, S. E. (2018). "Coronary artery calcification in patients diagnosed with severe mental illness."

Kuhn, W. F., et al. (1988). "Psychopathology in heart transplant candidates." *Journal of Heart Transplants,* 7, 223–226.

Ladouceur, A. M., Parmar, B., Weber, S. C., et al. (2020). "Clusters of bacterial RNA polymerase are biomolecular condensates that assemble through liquid-liquid phase separation." *bioRxiv.*

Laing, R. D. *The Facts of Life.* (New York: Pantheon Books, 1976), 34–46.

Lambert, N., Chen, Y. N., Cheng, Y. C., Li, C. M., Chen, G. Y., and Nori, F. (2013). "Quantum biology." *Nature Physics,* 9(1), 10.

Landau, Elizabeth (2020). "Mitochondria May Hold Keys to Anxiety and Mental Health." *Quanta Magazine.*

Landry, Susan H., Smith, Karen E., Swank, Paul R., Assel, Mike A., and Vellet, Sonya (2001). "Does early responsive parenting have a special importance for children's development or is consistency across early childhood necessary?" *Developmental Psychology*, 37(3), 387–403. http://dx.doi.org/10.1037/0012-1649.37.3.387.

Lane, Michelle, Robker, Rebecca L., and Robertson, Sarah A. (2014). "Parenting from before conception." *Science*, August 15, 2014, 756–760.

Lanphear, B. P. (2015). "The impact of toxins on the developing brain." *Annual Review of Public Health,* 36, 211–230.

Laplante, D. P., Barr, R. G., Brunet, A., Du Fort, G. G., Meaney, M. L., Saucier, J. F., and King, S. (2004). "Stress during pregnancy affects general intellectual and language functioning in human toddlers." *Pediatric Research*, 56(3), 400–410.

Laplante, D. P., Brunet, A., Schmitz, N., Chiampi, A., and King, S. (2008). "Project Ice Storm: Prenatal maternal stress affects cognitive and linguistic functioning in 5½-year-old children." *Journal of the American Academy of Child and Adolescent Psychiatry,* 47(9), 1063–1072.

Lauretti, E., Iuliano, L., and Praticò, D. (2017). "Extra-virgin olive oil ameliorates cognition and neuropathology of the 3xTg mice: role of autophagy." *Annals of Clinical and Translational Neurology*, 4(8), 564–574.

Lausten-Thomsen, U., Bille, D. S., Nässlund, I., Folskov, L., Larsen, T., Holm, J. C. (2013). "Neonatal anthropometrics and correlation to childhood obesity—data from the Danish Children's Obesity Clinic." *European Journal of Pediatrics*, 172(6), 747–751.

Lázaro, J., Dechmann, D. K., LaPoint, S., Wikelski, M., and Hertel, M. (2017). "Profound reversible seasonal changes of individual skull size in a mammal." *Current Biology,* 27(20), R1106-R1107.

Lebel, Catherine, Walton, Matthew Dewey, et al. (2016). "Prepartum and Postpartum Maternal Depressive Symptoms Are Related to Children's Brain Structure in Preschool." *Biological Psychiatry*, 2016, 80(11): 859. doi: 10.1016/j.biopsych.2015.12.004.

Leckman, J. F. and March, J. S. (2011). "Developmental neuroscience comes of age" (editorial). *Journal of Child Psychology and Psychiatry,* 52, 333–338.

Lee, H. S., Ghetti, A., Pinto-Duarte, A., Wang, X., Dziewczapolski, G., Galimi, F., and Sejnowski, T. J. (2014). "Astrocytes contribute to gamma oscillations and recognition memory." Proceedings of the National Academy of Sciences, 111(32), E3343–E3352.

Lee, Hojun, Kim, Boa, Kawata, Keisuke, et al. (2015). "Cellular Mechanism of Muscle Memory: Effects on Mitochondrial Remodeling and Muscle Hypertrophy." *Medicine and Science in Sports and Exercise,* 47(5S), 101–102.

Levin, M. (2012). "Molecular bioelectricity in developmental biology: New tools and recent discoveries: control of cell behavior and pattern formation by transmembrane potential gradients." *Bioessays*, 34, 205–217.

Levin, M. (2013). "Reprogramming cells and tissue patterning via bioelectrical pathways: molecular mechanisms and biomedical opportunities." Wiley Interdisciplinary Reviews: *Systems Biology and Medicine*, 5, 657–676.

Levin, M. and Stevenson, C. G. (2012). "Regulation of cell behavior and tissue patterning by bioelectrical signals: Challenges and opportunities for biomedical engineering." *Annual Review of Biomedical Engineering*, 14, 295–323.

Levin, M. (2013). "Remembrance of Brains Past." http://thenode.biologists.com/remembrance-of-brains-past/research/.

Lewin, R. (1980). "Is your brain really necessary?" *Science,* 210(4475), 1232–1234, 10.1126/science.6107993.

Li, D., Liu, L., and Odouli, R. (2009). "Presence of depressive symptoms during early pregnancy and the risk of preterm delivery: A prospective cohort study." *Human Reproduction*, 24(1), 146–153.

Libet, B. (1985). "Unconscious cerebral initiative and the role of conscious will in voluntary action." *Behavioral and Brain Sciences*, 8(4), 529–39.

Liester, M. and Liester, M. (2019). "Personality changes following heart transplants: Can epigenetics explain these transformations?" *European Psychiatry*, 56, S568–S568.

Liester, Mitchell and Liester, Maya. (2019). "A Retrospective Phenomenological Review of Acquired Personality Traits Following Heart Transplantation." April 2019. Conference: 27th European Congress of Psychiatry At: Warsaw, Poland.

Lim, K., Hyun, Y.-M., Kim, M., et al. (2015). "Neutrophil trails guide influenza-specific CD8 T cells in the airways." *Science*, 349(6252), aaa4352.

Lipton, B. H. (2001). "Insight into cellular 'consciousness.'" *Bridges*, 12(1), 5.

Lipton, B. H. *The Biology of Belief.* (Santa Rosa, Calif.: Mountain Of Love/Elite Books, 2005).

Liu, Jianghong, Raine, Adrian, Mednick, Sarnoff, et al. (2009). "The Association of Birth Complications and Externalizing Behavior in Early Adolescents: Direct and Mediating Effects." *Journal of Research on Adolescents*, 19(1), 93–111.

Liu, Xu, Ramirez, Steve, Tonegawa, Susumu, et al. (2012). "Optogenetic stimulation of a hippocampal engram activates fear memory recall." *Nature*, 484, 381–385 doi:10.1038/nature11028.

Livet, J., Weissman, T. A., Kang, H., Draft, R. W., Lu, J., Bennis, R. A., Sanes, J. R., and Lichtman, J. W. (2007). "Transgenic strategies for combinatorial expression of fluorescent proteins in the nervous system." *Nature*, 1, 450(7166), 56–62.

Lloyd, S. (October 2011). "Quantum coherence in biological systems." *Journal of Physics-Conference Series*, 302(1), 12037.

Lobo, D., Beane, W. S., and Levin, M. (2012) "Modeling planarian regeneration: A primer for reverse-engineering the worm." *PLoS Computational Biology*, 8(4), e1002481.

Lorber, J. (1978). "Is Your Brain Really Necessary?" *Archives of Disease in Childhood*, 53(10), 834–835.

Lowdin, P. O. "Quantum genetics and the aperiodic solid. Some aspects on the Biological problems of heredity, mutations, aging and tumors in view of the quantum theory of the DNA molecule." *Advances in Quantum Chemistry*, 2, 213–360. (Cambridge, Mass.: Academic Press, 1965).

Lunde, D. T. (1967). "Psychiatric complications of heart transplants." *American Journal of Psychiatry*, 124, 1190–1195.

Lutchmaya, S., Baron-Cohen, S., and Raggart, P. (2002). "Human sex differences in social and non-social looking preferences at 12 months of age." *Infant Behavior and Development*, 25, 319–325.

Ma, X. S., Zotter, S., Kofler, J., Ursin, R., Jennewein, T., Brukner, Č., and Zeilinger, A. (2012). "Experimental delayed-choice entanglement swapping." *Nature Physics*, 8(6), 479.

Mahler, M. S. *On Human Symbiosis and the Vicissitudes of Individuation.* Infantile Psychosis, vol. 1. (New York: International Universities Press, 1968).

Mai, F. M. (1986). "Graft and donor denial in heart transplant recipients." *American Journal of Psychiatry*, 143, 1159–1161.

Majorek, M. B. (2012). "Does the brain cause conscious experience?" *Journal of Consciousness Studies*, 19, 121–144.

Malaspina, Dolores, et al. (2008). "Stress During Pregnancy May Predispose to Schizophrenia." In John M. Grohol, *Psych Central.* August 21, 2008.

Marc Joanisse in Semeniuk, Ivan (2016). "New brain map reveals a world of meaning." *Globe and Mail.*

Marek, S., Siegel, J. S., Gordon, E. M., Raut, R. V., Gratton, C., Newbold, D. J., and Zheng, A. (2018). "Spatial and temporal organization of the individual human cerebellum." *Neuron*, 100(4), 977–993.

Markovich, Matt (2015). "Blow to the head turns Tacoma man into a genius." *Komono News*.

Marsh, Abigail A., Blair, R. J. R., et al. (2008). "Reduced Amygdala Response to Fearful Expressions in Children and Adolescents With Callous-Unemotional Traits and Disruptive Behavior Disorders." *American Journal of Psychiatry*, 165, 712–720.

Marston, D. J., Anderson, K. L., Hanein, D., et al. (2019). "High Rac1 activity is functionally translated into cytosolic structures with unique nanoscale cytoskeletal architecture." *Proceedings of the National Academy of Sciences*, 116(4), 1267–1272.

Martin, S. J., Grimwood, P. D., Morris, R.G.M. (2000). "Synaptic plasticity and memory: An evaluation of the hypothesis." *Annual Review Of Neuroscience*, 23, 649–711.

Mathis, A., Ferrari, M. C., Windel, N., Messier, F., Chivers, D. P. (2008). "Learning by embryos and the ghost of predation future." *Proceedings Biological Sciences*, 275, 2603–2607.

Matsumoto, Y. and Mizunami, M. (2002) "Lifetime olfactory memory in the cricket *Gryllus bimaculatus*." *Journal of Comparative Physiology: A-Neuroethology Sensory Neural and Behavioral Physiology*, 188, 295–299.

Mayo Clinic. http://www.mayoclinic.org/drugs-supplements/dhea/background/HRB-2005 9173.

McCaig, C. D., Rajnicek, A. M., Song, B., and Zhao, M. (2005). "Controlling cell behavior electrically: current views and future potential." *Physiological Reviews*, 85(3), 943–78.

McCarthy, Deirdre M., Morgan, Thomas J., Bhide, Pradeep G., et al. (2018). "Nicotine exposure of male mice produces behavioral impairment in multiple generations of descendants." *PLoS Biology*, 16(10), e2006497.

McClelland III, S. and Maxwell, R. E. (2007). "Hemispherectomy for intractable epilepsy in adults: The first reported series." *Annals of Neurology*, 61(4), 372–376.

McConnell, J. V., Jacobson, A. L., and Kimble, D. P. (1959). "The effects of regeneration upon retention of a conditioned response in the planarian." *Journal of Comparative and Physiological Psychology*, 52, 1–5.

McCraty, R. (2000) "Psychophysiological coherence: A link between positive emotions, stress reduction, performance and health." Proceedings of the Eleventh International Congress on Stress, Mauna Lani Bay, Hawaii.

McCraty, R. "Heart–brain neurodynamics: The making of emotions." (Publication No. 03-015). Boulder Creek, Calif.: HeartMath Research Center, 2003, Institute of HeartMath.

McCraty, R., and Tomasino, D. "The coherent heart: Heart-brain interactions, psychophysiological coherence, and the emergence of system wide order." (Publication No. 06- 022). (Boulder Creek, Calif.: HeartMath Research Center, 2006). Institute of HeartMath.

McDonnell Genome Institute, http://genome.wustl.edu/genomes/detail/physarum-polycephalum/.

McGill University. (2008). "Breastfeeding Associated With Increased Intelligence, Study Suggests." *ScienceDaily*; and Mortensen, E. L., Michaelsen, K. F., Sanders, S. A., and Reinisch, J. M. (2002). "The association between duration of breastfeeding and adult intelligence." *Journal of the American Medical Association*, 8, 287(18), 2365–2371.

McGowan, P. O. and Szyf, M. (2010). "The epigenetics of social adversity in early life: Implications for mental health outcomes." *Neurobiology of Disease*, 39(1), 66–72.

McGowan, P. O., Meaney, M. J., and Szyf, M. (2008). "Diet and the epigenetic (re)programming of phenotypic differences in behavior." *Brain Research*, 1237, 12–24.

Meaney, M. J. (2001). "Maternal care, gene expression, and the transmission of individual differences in stress reactivity across generations." *Annual Review of Neuroscience*, 24, 1161–1192.

MedicineNet. Medical Definition of Placebo Effect. Retrieved from: medicinenet.com/script/main /art.asp?articlekey=31481.

Merck Manual. Retrieved from http://www.merckmanuals.com/en-ca/home/heart-and -blood-vessel-disorders/biology-of-the-heart-and-blood-vessels/biology-of-the-heart.

Mesman. J., van IJzendoorn, M. H., Bakermans-Kranenburg, M. J. (2011). "Unequal in opportunity, equal in process: Parental sensitivity promotes child development in ethnic minority families." *Child Development Perspectives.* doi: 10.1111/j.1750-8606.2011.00223.x.

Metamorphosis: https://www.merriam-webster.com/dictionary/metamorphosis.

Mews, Philipp, Donahue, Greg, Berger, Shelley L., et al. (2017). "Acetyl-CoA synthetase regulates histone acetylation and hippocampal memory." *Nature.*

Meyer, K., Köster, T., Nolte, C., Weinholdt, C., Lewinski, M., Grosse, I., and Staiger, D. (2017). "Adaptation of iCLIP to plants determines the binding landscape of the clock-regulated RNA-binding protein At GRP7." *Genome Biology*, 18(1), 204.

Miller, A. H. and Raison, C. L. (2015). "The role of inflammation in depression: from evolutionary imperative to modern treatment target." *Nature Reviews Immunology*, 16(1), 22–34.

Millesi, E., Prossinger, H., Dittami, J. P., and Fieder, M. (2001). "Hibernation effects on memory in European ground squirrels (Spermophilus citellus)." *Journal of Biological Rhythms*, 16, 264–271.

Morsella, E., Godwin, C. A., Gazzaley, A., et al. (2016). "Homing in on consciousness in the nervous system: An action-based synthesis." *Behavioral and Brain Sciences*, 39.

Moelling, Karin (2012). "Are viruses our oldest ancestors?" *EMBO Reports*, 13(12), 1033.

Montiel-Castro, A. J., González-Cervantes, R. M., Bravo-Ruiseco, G., and Pacheco-López, G. (2013). "The microbiota-gut-brain axis: neurobehavioral correlates, health and sociality." *Frontiers in Integrative Neuroscience*, 7.

Moore, R. S., Kaletsky, R., and Murphy, C. T. (2019). "Piwi/PRG-1 argonaute and TGF-β mediate transgenerational learned pathogenic avoidance." *Cell*, 177(7), 1827–1841.

Morgan, Christopher P. and Bale, Tracy L. (2011). "Early prenatal stress epigenetically programs dysmasculinization in second-generation offspring via the paternal lineage." *Journal of Neuroscience*, 17; 31(33), 11748–11755.

Morgan, N., Irwin, M. R., Chung, M., and Wang, C. (2014). "The effects of mind-body therapies on the immune system: meta-analysis." *PLoS One*, 9(7), e100903.

Moran, I., Nguyen, A., Munier, C.M.L., et al. (2018). "Memory B cells are reactivated in subcapsular proliferative foci of lymph nodes." *Nature Communications*, 9(1), 1–14.

Muckli, Lars (2009). "Scientists reveal secret of girl with 'all seeing eye.'" http://www.gla.ac.uk/news/archiveofnews/2009/july/headline_125704_en.html.

Mudd, A. T., Berding, K., Wang, M., Donovan, S. M., and Dilger, R. N. (2017). "Serum cortisol mediates the relationship between fecal Ruminococcus and brain N-acetylaspartate in the young pig." *Gut Microbes*, 1–12.

Muenke, Max (2007). "Tiny brain no obstacle to French civil servant." *Reuters Health News.*

Muller, V. and Lindenberger, U. (2011) "Cardiac and Respiratory Patterns Synchronize between Persons during Choir Singing." *PLoS One*, 6(9), e24893.

Murillo, O. D., Thistlethwaite, W., Kitchen, R. R., et al. (2019). "exRNA atlas analysis reveals distinct extracellular RNA cargo types and their carriers present across human biofluids." *Cell*, 177(2), 463–477.

Mustard, J. and Levin, M. (2014). "Bioelectrical mechanisms for programming growth and form: taming physiological networks for soft body robotics." *Soft Robotics*, 1(3), 169–191.

Myhrer, T. (2003). "Neurotransmitter systems involved in learning and memory in the rat: a meta-analysis based on studies of four behavioral tasks." *Brain Research Brain Research Reviews*, 41(2–3):268–287.

Naik, S., Larsen, S. B., Fuchs, E., et al. (2017). "Inflammatory memory sensitizes skin epithelial stem cells to tissue damage." *Nature*, 550(7677), 475–480.

Naik, S., Larsen, S. B., Cowley, C. J., and Fuchs, E. (2018). "Two to tango: Dialog between immunity and stem cells in health and disease." *Cell*, 175(4), 908–920.

Nakagaki, T., Kobayashi, R., Nishiura, Y., and Ueda, T. (2004). "Obtaining multiple separate food sources: Behavioral intelligence in the Physarum plasmodium." *Royal Society*, 271(1554).

Nakagaki, T., Yamada, H., and Hara, M. (2004). "Smart network solutions in an amoeboid organism." *Biophysical Chemistry*, 107(1), 1–5.

Nakagaki, T., Yamada, H., and Tóth, Á. (2000). "Intelligence: Maze-solving by an amoeboid organism." *Nature*, 407(6803), 470.

Nasrollahi, S., Walter, C., and Pathak, A., et al. (2017). "Past matrix stiffness primes epithelial cells and regulates their future collective migration through a mechanical memory." *Biomaterials*, 146, 146–155.

Nguyen, T. T., Zhang, X., Wu, T. C., Liu, J., Le, C., Tu, X. M., and Jeste, D. V. (2021). "Association of Loneliness and Wisdom with Gut Microbial Diversity and Composition: An Exploratory Study." *Frontiers in Psychiatry*, 12, 395.

Nie, Duyu, Di Nardo, Alessia, Sahin, Mustafa, et al. (2010). "Tsc2-Rheb signaling regulates EphA-mediated axon guidance." *Nature Neuroscience*, 13, 163–172. doi:10.1038/nn.2477.

Nili, Hussein, Walia, Sumeet, Sriram, Sharath, et al. (2015). "Donor-Induced Performance Tuning of Amorphous SrTiO3Memristive Nanodevices: Multistate Resistive Switching and Mechanical Tunability." *Advanced Functional Materials*, 25(21), 3172–3182.

Nobel Prize in Medicine, 2017. https://www.nobelprize.org/nobel_prizes/medicine/laureates/2017/.

Nordenfelt, P., Elliott, H. L., and Springer, T. A. (2016). "Coordinated integrin activation by actin-dependent force during T-cell migration." *Nature Communications*, 7, 13119.

Noreen, Faiza, Röösli, Martin, Gaj, Pawel, et al. (2014). "Modulation of Age- and Cancer-Associated DNA Methylation Change in the Healthy Colon by Aspirin and Lifestyle." *Journal of the National Cancer Institute*, 106(7).

Northstone, K., Lewcock, M., Groom, A., Boyd, A., Macleod, J., Timpson, N., and Wells, N. (2019). "The Avon Longitudinal Study of Parents and Children (ALSPAC): an update on the enrolled sample of index children in 2019." *Wellcome open research*, 4.

Nuñez, J. K., Lee, A. S., Engelman, A., and Doudna, J. A. (2015). "Integrase-mediated spacer acquisition during CRISPR-Cas adaptive immunity." *Nature*, 519(7542), 193.

O'Donnell, Kieran J., Glover, Vivette, O'Connor, Thomas G., et al. (2014). "The persisting effect of maternal mood in pregnancy on childhood psychopathology," *Development and Psychopathology*, 26(2), 393–403. doi: https://doi.org/10.1017/S0954579414000029.

Ogawa, T. and de Bold, A. J. (2014). "The heart as an endocrine organ." *Endocrine Connections*, 3(2), R31–R44.

Ogbonnaya, E. S., Clarke, G., Shanahan, F., Dinan, T. G., Cryan, J. F., and O'Leary, O. F. (2015). "Adult hippocampal neurogenesis is regulated by the microbiome." *Biological Psychiatry*, 78(4), e7–e9.

Ohio State University. (November 6, 2018). "Immune system and postpartum depression linked? Research in rats shows inflammation in brain region after stress during pregnancy." *ScienceDaily*.

Olender, T., Lancet, D., Nebert, D. W. (2008–2009). "Update on the olfactory receptor (or) gene superfamily." *Human Genomics*, 3, 87–97.

Olson, J. A., Suissa-Rocheleau, L., Lifshitz, M., Raz, A., and Veissiere, S. P. (2020). "Tripping on nothing: Placebo psychedelics and contextual factors." *Psychopharmacology*, 1–12.

Oofana, Ben (April 15, 2018). "Dissolving the Layers of Emotional Body Armor." https://benoofana.com/dissolving-the-layers-of-emotional-body-armor.

Open Sciences, "Manifesto for a Post-Materialist Science." *Open Sciences*, http://opensciences.org/about/manifesto-for-a-post-materialist-science.

Ornish, D. *Love and Survival: The Scientific Basis for the Healing Power of Intimacy* (New York: HarperCollins, 1998).

Ossola, Paolo, Garrett, Neil, Sharot, Tali, Marchesi, Carlo (2020). "Belief updating in bipolar disorder predicts time of recurrence." *eLife*, 9.

Paanksepp, J. *Affective Neuroscience*. (London: Oxford University Press, 2004).

Pace, T.W.D., Cole, S. P., Raison, C. L., et al. (2009). "Effect of compassion meditation on neuroendocrine, innate immune and behavioral responses to psychosocial stress." *Psychoneuroendocrinology*, 34(1), 87–98.

Pal, D., Dean, J. G., and Hudetz, A. G., et al. (2018). "Differential role of prefrontal and parietal cortices in controlling level of consciousness." *Current Biology*, 28, 2145.e5–2152.e5. doi: 10.1016/j.cub.2018.05.025.

Parnia, S., Spearpoint, K., and Wood, M., et al. (2014). "AWARE—AWAreness during REsuscitation—A prospective study." *Resuscitation*, 85(12), 1799–1805.

Pearsall, P., Schwartz, G. E., and Russek, L. G. (2002). "Changes in heart transplant recipients that parallel the personalities of their donors." *Journal of Near-Death Studies*, 20(3), 191–206.

Pearsall, Paul, Schwartz, Gary E., and Russek, Linda G. (2005). "Organ Transplants and Cellular Memories." *Nexus Magazine*, 12(3).

Pearce, K., Cai, D., Roberts, A. C., et al. (2017). "Role of protein synthesis and DNAmethylation in the consolidation and maintenance of long-term memory in Aplysia." *eLife*, 6.

Pearson, Kevin (2016). "Exercise during pregnancy may reduce markers of aging in offspring." ScienceDaily. www.sciencedaily.com/releases/2016/11/161104120458.htm.

Pembrey, M., Saffery, R. and Bygren, L. O. (2015). "Network in Epigenetic Epidemiology. Human transgenerational responses to early-life experience: Potential impact on development, health and biomedical research." *Journal of Medical Genetics*, 51, 563–72.

Penrose, R. *The Emperor's New Mind*. (Oxford, UK: Oxford University Press, 1989).

Penrose, R. *Shadows of the Mind: A Search for the Missing Science of Consciousness*, (Oxford, UK: Oxford University Press, 1994).

Perera, F. P., Li, Z., Whyatt, R., Hoepner, L., Wang, S., Camann, D., and Rauh, V. (2009). "Prenatal airborne polycyclic aromatic hydrocarbon exposure and child IQ at age 5 years." *Pediatrics*, 124(2), e195–e202.

Perone, S., Almy, B., and Zelazo, P. D. "Toward an understanding of the neural basis of executive function development." In *The Neurobiology of Brain and Behavioral Development* (Cambridge, Mass.: Academic Press, 2018), 291–314.

Persico, M., Podoshin, L., Fradis, M., Golan, D., Wellisch, G. (1983). "Recurrent middle-ear infections in infants: The protective role of maternal breast feeding." *Ear, Nose, & Throat Journal*, 62(6), 297–304.

Peterson, Eric (2018). "Stem Cells Remember Tissues' Past Injuries." *Quanta Magazine*.

Pfeiffer, Ronald F. (2007). "Neurology of Gastroenterology and Hepatology." In *Neurology and Clinical Neuroscience*, 1511–1524 m. *Biosystems*, 112(1), 1–10.

Physicsworld.com. (2016). "Double quantum-teleportation milestone is Physics World 2015 Breakthrough of the Year." physicsworld.com.

Pitman, Teresa. (September 7, 2016) "The dos and don'ts of safe formula feeding." Retrieved from https://www.todaysparent.com/baby/baby-food/the-dos-and-donts-of-safe-formula-feeding/.

Plasma cell. (July 4, 2017). In *Wikipedia*. Retrieved 17:16, September 13, 2017, from https://en.wikipedia.org/w/index.php?title=Plasma_cell&oldid=788883461).

Playfair, Guy Lyon (2003). *Twin Telepathy: The Psychic Connection*. (New York: Vega Books, 2003).

Poo, M. M., Pignatelli, M., Ryan, T. J., Tonegawa, S., Bonhoeffer, T., Martin, K. C., et al. (2016). "What is memory? The present state of the engram." *BMC Biology*, 14, 40.

Popkin, Gabriel (2017). "Bacteria Use Bursts of Electricity to Communicate." *Quanta Magazine*.

Popper, K. and Eccles, J. C. *The Self-Conscious Mind and the Brain. In: The Self and Its Brain*. (London: Routledge, 2000), 355–376.

Posner, R., Toker, I. A., Antonova, O., Star, E., Anava, S., Azmon, E., and Rechavi, O. (2019). "Neuronal small RNAs control behavior transgenerationally." *Cell*, 177(7), 1814–1826.

Post, C. M., Boule, L. A., Burke, C. G., O'Dell, C. T., Winans, B., and Lawrence, B. P. (2019). "The ancestral environment shapes antiviral CD8+ T cell responses across generations." *iScience*, 20, 168–183.

"Post-Materialist Science." Open Sciences, http://opensciences.org/about/manifesto-for-a-post-materialist-science.

Potter, Garrett D., Byrd, Tommy A., Mugler, Andrew, and Sun, Bo (2016). "Communication shapes sensory response in multicellular networks." *Proceedings of the National Academy of Sciences*, 113(37), 10334–10339.

Poulopoulos, A., Murphy, A. J., Macklis, J. D., et al. (2019). "Subcellular transcriptomes and proteomes of developing axon projections in the cerebral cortex." *Nature*, 565(7739), 356–360.

Powers, Jenny (2018). "Increased Risk of Mortality From Heart Disease in Patients With Schizophrenia." Presented at EPA. https://www.firstwordpharma.com/node/1547343.

Pribram, K. H. (2012): "The Implicate Brain." In B. Hiley and F. D. Peat (eds.), *Quantum Implications: Essays in Honour of David Bohm*. (New York: Routledge, 1987), 365–371.

Prindle, A., Liu, J., Asally, M., Ly, S., Garcia-Ojalvo, J., and Süel, G. M. (2015). "Ion channels enable electrical communication in bacterial communities." *Nature*, 527(7576), 59–63.

Pryce, C. R. and Feldon, J. (2003). "Long-term neurobehavioral impact of postnatal environment in rats: Manipulations, effects and mediating mechanisms." *Neuroscience & Biobehavioral Reviews*, 27, 57–71.

Pseudopodium: https://www.merriam-webster.com/dictionary/pseudopodium.

Qin, J., Li, R., and Mende, D. R., et al. (2010). "A human gut microbial gene catalogue established by metagenomic sequencing." *Nature*, 464(7285), 59–65.

quorum sensing: https://en.wikipedia.org/wiki/Quorum_sensing.

Raby, K. Lee; Roisman, Glenn I. R.; Simpson, Jeffry A., et al. (2015). "The Enduring Predictive Significance of Early Maternal Sensitivity: Social and Academic Competence Through Age 32 Years." *Child Development*, 86(3), 695–708. doi: 10.1111/cdev.12325.

Radin, D. *Entangled Minds*. (New York: Pocket Books, 2006); Kauffman, S. (2010). "Is There A 'Poised Realm' Between the Quantum and Classical Worlds?" *Cosmos and Culture*.

Radin, D. I., Taft, R., and Yount, G. (2004). "Possible effects of healing intention on cell cultures and truly random events." *Journal of Alternative and Complementary Medicine*, 10, 103–112.

Rai, Mamta, Coleman, Zane, and Demontis, Fabio, et al. (2021). "Proteasome stress in skeletal muscle mounts a long-range protective response that delays retinal and brain aging." *Cell Metabolism*.

Raine, Adrian, Brennan, P., Mednick, S. A. (1994). "Birth Complications Combined With Early Maternal Rejection at Age 1 Year Predispose to Violent Crime at Age 18 Years." *Archives of General Psychiatry*, 51, 948–988.

Raine, Adrian, Brennan, Patricia, and Mednick, S. A. (1997). "Interaction Between Birth Complications and Early Maternal Rejection in Predisposing Individuals to Adult Violence: Specificity to Serious, Early-Onset Violence." *American Journal of Psychiatry*, 154, 1265–1271.

Raison, C. L. and Miller, A. H. (2013). "The evolutionary significance of depression in Pathogen Host Defense (PATHOS-D)." *Molecular Psychiatry*, 18, 15–37.

Ramilowski, J. A., Goldberg, T., Harshbarger, J., Kloppman, E., Lizio, M., Satagopam, V. P., and Forrest, A. R. (2015). "A draft network of ligand-receptor-mediated multicellular signaling in human. *Nature Communications*, 6.

Rantakallio P., Laara, E., Isohanni, and M., Moilanen, I. (1992). "Maternal smoking during pregnancy and delinquency of the offspring: an association without causation?" *International Journal of Epidemiology*, 21(6), 1106–1113.

Rao, G. and Rowland, K. (2011). "Zinc for the common cold—not if, but when." *The Journal of Family Practice*, 60(11), 669.

Rao, M. and Gershon, M. D. (2017). "The dynamic cycle of life in the enteric nervous system." *Nature Reviews Gastroenterology & Hepatology*, 14(8), 453–454.

Rasmussen, H. N., Scheier, M. F., and Greenhouse, J. B. (2009). "Optimism and physical health: A meta-analytic review." *Annals of Behavioral Medicine*, 37(3), 239–256.

Ray, S. (1999) "Survival of olfactory memory through metamorphosis in the fly Musca domestica." *Neuroscience Letters*, 259, 37–40.

Rechavi, O., Minevich, G., and Hobert, O. (2011). "Transgenerational inheritance of an acquired small RNA-based antiviral response in C. elegans." *Cell*, 147(6), 1248–1256.

Reddivari, Lavanya, Veeramachaneni, D. N., Rao, Vanamala, Jairam K. P., et al. (2017). Perinatal Bisphenol A Exposure Induces Chronic Inflammation in Rabbit Offspring via Modulation of Gut Bacteria and Their Metabolites." *mSystems*, 2(5).

Reid, C. R., MacDonald, H., Mann, R. P., Marshall, J. A., Latty, T., and Garnier, S. (2016). "Decision-making without a brain: How an amoeboid organism solves the two-armed bandit." *Journal of The Royal Society Interface*, 13(119), 20160030.

Renken, Elena (2019). "How Microbiomes Affect Fear." *Quanta Magazine*.

Rennie, John and Reading-Ikkanda, Lucy (2017) "Seeing the Beautiful Intelligence of Microbes." *Quanta Magazine*.

Ressem, Synnøve (2010). "Inside a moth's brain." *Gemini Magazine*.

Ribosome: In *Wikipedia*. Retrieved from https://en.wikipedia.org/wiki/Ribosome.

Rice University (August 26, 2019). "Scientists advance search for memory's molecular roots: Architecture of the cytoskeleton in neurons." ScienceDaily. Retrieved November 22, 2020, from www.sciencedaily.com/releases/2019/08/190826150658.htm.

Rietdorf, K. and Steidle, J.L.M. (2002) "Was Hopkins right? Influence of larval and early adult experience on the olfactory response in the granary weevil. *Sitophilus granaries* (Coleoptera, Curculionidae)." *Physiological Entomology*, 27, 223–227.

Rimer, S. and Drexler, M. (2011). "The biology of emotion and what it may teach us about helping people to live longer." *Harvard Public Health Review*, Winter, 813. Retrieved from https://www.hsph.harvard.edu/news/magazine/happiness-stress-heart-disease/.

Rimm, E. B., Appel, L. J., Lichtenstein, A. H., et al. (2018). "Seafood long-chain n-3 polyunsaturated fatty acids and cardiovascular disease: A science advisory from the American Heart Association." *Circulation*, 138(1), e35–e47.

Risch, N. J. (2000). "Searching for genetic determinants in the new millennium." *Nature*, 405(6788), 847–856.

Robert Malenka in Fields, R. Douglas (2013). "Human Brain Cells Make Mice Smart." *Scientific American*.

Robert Sternberg in *Tufts University News* (2008). Biologist Michael Levin Joins Tufts University [news release]. Retrieved from http://now.tufts.edu/news-releases/biologist-michael-levin-joins-tufts-university.

Robins, J.L.W., McCain, N. L., Gray, D. P., Elswick, R. K., Walter, J. M., and McDade, E. (2006). "Research on psychoneuroimmunology: tai chi as a stress management approach for individuals with HIV disease." *Applied Nursing Research*, 19(1), 2–9. [PubMed: 16455435]

Rodgers, Paul (2014). "Einstein Was Right: You Can Turn Energy Into Matter." https://www.forbes.com/sites/paulrodgers/2014/05/19/einstein-was-right-you-can-turn-energy-into-matter/?sh=2c58c85126ac.

Rosenbaum, M. B., Sicouri, S. J., Davidenko, J. M., and Elizari, M. V. (1985). "Heart rate and electrotonic modulation of the T wave: a singular relationship." *Cardiac Electrophysiology and Arrhythmias*. Grune and Stratton.

Roth, T. L., Lubin, F. D., Sodhi, M., Kleinman, J. E. (2009). "Epigenetic mechanisms in schizophrenia." *Biochimica et Biophysica Acta (BBA) - General Subjects*; 1790:869–877.

Roth, Tania L. and Sweatt, J. David (2011). "Epigenetic mechanisms and environmental shaping of the brain during sensitive periods of development." *Journal of Child Psychology and Psychiatry*, 52(4), 398–408. doi:10.1111/j.1469-7610.2010.02282.x.

Roth, Tania L.; Lubin, F.; Funk, Adam J.; Sweatt, J. David. (2009). "Lasting Epigenetic Influence of Early-Life Adversity on the *BDNF* Gene." *j.biopsych*, 65(9), 760–769.

Roth, Tania L. and Sweatt , J. David (2011). "Epigenetic marking of the BDNF gene by early-life adverse experiences." *Hormones and Behavior*, Special Issue: Behavioral Epigenetics, 59(3), 315–320.

Rowland, Katherine (2018). "We Are Multitudes." *Aeon Magazine*.

Ruczynski, I. and Siemers, B. M. (2011). "Hibernation does not affect memory retention in bats." *Biology Letters*, 7(1), 153–155.

Russell, W. R., Hoyles, L., Flint, H. J., and Dumas, M. E. (2013). "Colonic bacterial metabolites and human health." *Current Opinion in Microbiology*, 16(3), 246–254.

Rutherford, O. M. and Jones, D. A. (1986). "The role of learning and coordination in strength training." *European Journal of Applied Physiology and Occupational Physiology*, 55, 100–105.

Ryan, T. J., Roy, D. S., Pignatelli, M., Arons, A., and Tonegawa, S. (2015). "Engram cells retain memory under retrograde amnesia." *Science*, 348, 1007–1013.

Sadanand, Fulzele, Bikash, Sahay, Carlos, M. Isales, et al. (2020). "COVID-19 Virulence in Aged Patients Might Be Impacted by the Host Cellular MicroRNAs Abundance/Profile." *Aging and Disease*, 11(3), 509–522.

Sahu, S., Ghosh, S., Hirata, K., Fujita, D., and Bandyopadhyay, A. (2013). "Multi-level memory-switching properties of a single brain microtubule." *Applied Physics Letters*, 102(12), 123701.

Saigle, Victoria, Dubljević, Veljko, and Racine, Eric (2018). "The Impact of a Landmark Neuroscience Study on Free Will: A Qualitative Analysis of Articles Using Libet and Colleagues' Methods." *AJOB Neuroscience*, 9(1), 29.

Saigusa, T., Tero, A., Nakagaki, T., and Kuramoto, Y. (2008). "Amoebae anticipate periodic events." *Physical Review Letters*, 100(1), 018101.

Salone, L. R., Vann, W. F., Dee, D. L. (2013). "Breastfeeding: An overview of oral and general health benefits." *Journal of the American Dental Association* (1939), 144(2), 143–151.

SAMHSA, Substance Abuse and Mental Health Services Administration. (2013). *Results from the 2012 National Survey on Drug Use and Health: Detailed Tables*. Rockville, Md.: SAMHSA, Center for Behavioral Health Statistics and Quality and Health Canada. (2013). *Canadian Alcohol and Drug Use Monitoring Survey: Summary of results for 2012*. Ottawa: http://www.hc-sc.gc.ca/hc-ps/drugsdrogues/stat/_2012/summary-sommaire-eng.php.

Sampson, T. R. and Mazmanian, S. K. (2015). "Control of brain development, function, and behaviour by the microbiome." *Cell Host & Microbe*, 17(5), 565–576.

Sadanand, Fulzele, Bikash, Sahay, Carlos, M. Isales, et al. (2020). "COVID-19 Virulence in Aged Patients Might Be Impacted by the Host Cellular MicroRNAs Abundance/Profile." *Aging and Disease*, 11(3), 509–522.

Sanchez, M. M. (2006). "The impact of early adverse care on HPA axis development: Nonhuman primate models." *Hormones and Behavior*, 50, 623–631. [PubMed: 16914153]

Sandman, C. A., Walker, B. B., and Berka, C. (1982). "Influence of afferent cardiovascular feedback on behavior and the cortical evoked potential." *Perspectives in Cardiovascular Medicine*.

Sapolsky, Robert. *Why Zebras Don't Get Ulcers*. (New York. Henry Holt & Company, 2004).

Sarapas, C., et al. (2011). "Genetic markers for PTSD risk and resilience among survivors of the World Trade Center attacks." *Disease Markers*, 30(2-3), 101–10.

Sarnat, H. B. and Netsky, M. G. (1985). "The brain of the planarian as the ancestor of the human brain." *Canadian Journal of Neurological Sciences*, 12(4), 296–302.

Scheinin, A., Kallionpää, R. E., Revonsuo, A., et al. (2018). "Differentiating drug-related and state-related effects of dexmedetomidine and propofol on the electroencephalogram." *Anesthesiology: The Journal of the American Society of Anesthesiologists*, 129(1), 22–36.

Schemann, M., Frieling, T., and Enck, P. (2020). "To learn, to remember, to forget—How smart is the gut?" *Acta Physiologica*, 228(1), e13296.

Schillinger, Liesl (2003). "Odorama." *New York Times.*

Schlinzig, T., Johansson, S., Gunnar, A., Ekström, T. J., and Norman, M. (2009). "Epigenetic modulation at birth-altered DNA-methylation in white blood cells after Caesarean section." *Acta Pædiatrica*. 98:1096–1099.

Schmidt, A. and Thews, G. "Autonomic Nervous System." In Janig, W., *Human Physiology* (2nd ed.). (New York: Springer-Verlag, 1989). 333–370.

Schore, A. N. (2002). "Dysregulation of the right brain: A fundamental mechanism of traumatic attachment and the psychopathogenesis of posttraumatic stress disorder." *Australian and New Zealand Journal of Psychiatry*, 36, 9–30. [PubMed: 11929435]

Schore, Allan (2008). "Modern attachment theory: The central role of affect regulation in development and treatment." *Clinical Social Work Journal*, 36, 9–20.

Schore, Allan (2017a). "All Our Sons: The Developmental Neurobiology and Neuroendocrinology of Boys at Risk." *Infant Mental Health Journal*, 38(1), 15–52.

Schore, Allan (2017b). "Modern Attachment Theory. Chapter in APA Handbook of trauma psychology." Steven N. Gold, editor in chief.

Schramm, J., Kuczaty, S., Sassen, R., Elger, C. E., and Von Lehe, M. (2012). "Pediatric functional hemispherectomy: Outcome in 92 patients." *Acta Neurochirurgica*, 154(11), 2017–2028.

Schrödinger, E. *What Is life? and, Mind and Matter.* (Cambridge, UK: Cambridge University Press, 1967).

Schrott, R., Acharya, K., Murphy, S. K., et al. (2020). "Cannabis use is associated with potentially heritable widespread changes in autism candidate gene DLGAP2 DNA methylation in sperm." *Epigenetics*, 15(1–2), 161–173.

Schwartz, Gary E. *The Energy Healing Experiments: Science Reveals Our Natural Power to Heal.* (New York: Simon & Schuster, 2008).

Schwartz, Gary E. *The Sacred Promise: How Science Is Discovering Spirit's Collaboration with Us in Our Daily Lives.* (New York: Simon & Schuster, 2011).

Schwartz, Gary E. and Simon, W. L. *The Afterlife Experiments: Breakthrough Scientific Evidence of Life after Death.* (New York: Simon & Schuster, 2002).

Sell, C. S. (2006). "On the unpredictability of odor." *Angewandte Chemie International Edition*, 45, 6254–6261.

Seth, Anil K. (2017). "The real problem: It looks like scientists and philosophers might have made consciousness far more mysterious than it needs to be." *Aeon Magazine.*

Seung, Sebastian. *Connectome: How the Brain's Wiring Makes Us Who We Are.* (New York: Mariner Books, 2013).

Shapiro, J. A. (1998). "Thinking about bacterial populations as multicellular organisms." *Annual Review of Microbiology*, 52, 81–104.

Shapiro, J. A. (2007). "Bacteria are small but not stupid: Cognition, natural genetic engineering and socio-bacteriology." *Studies in History and Philosophy of Science Part C: Studies in History and Philosophy of Biological and Biomedical Sciences*, 38(4), 807–819.

Sheiman, I. M. and Tiras, K. L. "Memory and morphogenesis in planaria and beetle." In C. I. Abramson, Z. P. Shuranova, and Y. M. Burmistrov (eds.) *Russian Contributions to Invertebrate Behavior.* (Westport, Conn.: Praeger, 1996).

Shelley L. Berger in public release: University of Pennsylvania (May 31, 2017). "Metabolic enzyme fuels molecular machinery of memory." https://www.pennmedicine.org/news/news-releases/2017/may/metabolic-enzyme-fuels-molecular-machinery-of-memory.

Shenderov, B. A. (2012). "Gut indigenous microbiota and epigenetics." *Microbial Ecology in Health and Disease*, 23(1), 17195.

Shipman, Seth L., Nivala, Jeff, Macklis, Jeffrey D., Church, and George M., (2017). "CRISPR-Cas encoding of a digital movie into the genomes of a population of living bacteria." *Nature*, nature23017.

Shirakawa, T. and Gunji, Y. P. (2007). "Emergence of morphological order in the network formation of Physarum polycephalum." *Biophysical Chemistry*, 128(2), 253–260.

Shomrat, T. and Levin, M. (2013). "An automated training paradigm reveals long-term memory in planarians and its persistence through head regeneration." *Journal of Experimental Biology*, 216(20), 3799–3810.

Siegel, D. J. (2001). "Toward an interpersonal neurobiology of the developing mind: Attachment relationships, 'mindsight,' and neural integration." *Infant Mental Health Journal*, 22(1–2), 67–94.

Siklenka, Keith, Erkek, Serap, Kimmins, Sarah, et al. (2015). "Disruption of histone methylation in developing sperm impairs offspring health transgenerationally." *Science*, 350(6), 261.

Slavich, G. M. and Cole, S. W. (2013). "The emerging field of human social genomics." *Clinical Psychological Science*, 1(3), 331–348.

Smith-Ferguson, J., Reid, C. R., Latty, T., and Beekman, M. (2017). "Hänsel, Gretel and the slime mould—How an external spatial memory aids navigation in complex environments." *Journal of Physics D: Applied Physics*, 50(41), 414003.

Smith, Anne (2016). "When Breastfeeding Doesn't Work Out." *Breastfeeding Basics*. Retrieved from https://www.breastfeedingbasics.com/articles/when-breastfeeding-doesnt-work-out.

Smith, L. K. and Wissel, E. F. (2019). "Microbes and the mind: How bacteria shape affect, neurological processes, cognition, social relationships, development, and pathology." *Perspectives on Psychological Science*, 14(3), 397–418.

Smith, L. M., Cloak, C. C., Poland, R. E., Torday, J., and Ross, M. G (2003). "Prenatal nicotine increases testosterone levels in the fetus and female offspring." *Nicotine & Tobacco Research*, 5(3), 369–374.

Sniekers, Suzanne, Stringer, Sven, Chabris, Christopher F., et al. (2017). "Genome-wide association meta-analysis of 78,308 individuals identifies new loci and genes influencing human intelligence." *Nature Genetics*.

Sommer, F. and Bäckhed, F. (2013). "The gut microbiota--masters of host development and physiology." *Nature Reviews Microbiology*, 11(4), 227.

Sorscher, N. and Cohen, L. J. (1997). "Trauma in children of Holocaust survivors: Transgenerational effects." *American Journal of Orthopsychiatry*, 67(3), 493.

Spielrein, Sabina (1912). "Destruction as a Cause of Coming into Being." *Jahrbuch fur psychoanalytische und psychopathologische Forschungen*, 4:465–503, Vienna.

Stacho, Martin, Herold, Christina, Güntürkün, Onur, et al. (2020). "A cortex-like canonical circuit in the avian forebrain." *Science*.

Stahl, F. W. (1988) "Bacterial genetics. A unicorn in the garden." *Nature*, 335, 112.

Staron, R. S., Leonardi, M. J., Karapondo, D. L., Malicky, E. S., Falkel, J. E., Hagerman, F. C., and Hikida, R. S. (1991). "Strength and skeletal muscle adaptations in heavy-resistance-trained women after detraining and retraining." *Journal of Applied Physiology*, 70, 631–640.

Stary, V., Pandey, R. V., Stary, G., et al. (2020). "A discrete subset of epigenetically primed human NK cells mediates antigen-specific immune responses." *Science Immunology*, 5(52).

Steffener, Jason, Habeck, Christian, Stern, Yaakov, et al. (2016). "Differences between chronological and brain age are related to education and self-reported physical activity." *Neurobiology of Aging*, 40, 138. doi.

Stevenson, Ian (1984). "American Children Who Claim To Remember Previous Lives." *Journal of Nervous and Mental Disease*, 171, 742–748.

Stewart, S., Rojas-Munoz, A., and Izpisua Belmonte, J. C. (2007). "Bioelectricity and epimorphic regeneration." *Bioessays*, 29, 1133–1137.

Stoodley, C. and Schmahmann, J. (2009). "Functional topography in the human cerebellum: A meta-analysis of neuroimaging studies." *NeuroImage*, 44(2), 489–501. doi:10.1016/j.neuroimage.2008.08.039

Striepens, N., Kendrick, K. M., et al. (2011). "Prosocial effects of oxytocin and clinical evidence for its therapeutic potential." *Frontiers in Neuroendocrinology*, 32(4), 426–450.

Suan, D., Nguyen, A., Moran, I., Bourne, K., Hermes, J. R., Arshi, M., and Kaplan, W. (2015). "T follicular helper cells have distinct modes of migration and molecular signatures in naive and memory immune responses." *Immunity*, 42(4), 704–718.

Sullivan, K. J. and Storey, K. B. (2012) "Environmental stress responsive expression of the gene li16 in *Rana sylvatica*, the freeze tolerant wood frog." *Cryobiology*, 64, 192–200.

Sullivan, Regina (2014). NYU Langone Medical Center/New York University School of Medicine. "Mother's soothing presence makes pain go away, changes gene activity in infant brain." *ScienceDaily*, November 18, 2014. www.sciencedaily.com/releases/2014/11/141118125432.htm.

Sylvia, Claire. *A Change of Heart*. (New York: Warner Books, 1998).

Synnøve, Ressem (2010). "Inside a moth's brain." *Gemini Magazine* in EARTH.

Taaffe, D. R. and Marcus, R. (1997). "Dynamic muscle strength alterations to detraining and retraining in elderly men." *Clinical Physiology*, 17, 311–324.

Tabas, A., Mihai, G., Kiebel, S., Trampel, R., and von Kriegstein, K. (2020). "Abstract rules drive adaptation in the subcortical sensory pathway." *eLife*, 9, e64501.

Tabuchi, T. M., Rechtsteiner, A., and Strome, S., et al. (2018). "Caenorhabditis elegans sperm carry a histone-based epigenetic memory of both spermatogenesis and oogenesis." *Nature Communications*, 9(1), 1–11.

Tan, J., McKenzie, C., Potamitis, M., Thorburn, A. N., Mackay, C. R., and Macia, L. (2014). "The role of short-chain fatty acids in health and disease." *Advances in Immunology*, 121(91), e119.

Taylor, David C. (1988). "Oedipus's Parents Were Child Abusers." *British Journal of Psychiatry*, 153, 561–563.

Ted Kaptchuk in "The Power of the Placebo Effect." https://www.health.harvard.edu/mental-health/the-power-of-the-placebo-effect#:~:text=%22The%20placebo%20effect%20is%20a,need%20the%20ritual%20of%20treatment.

Templeton, G. (2014). "Smart dust: A complete computer that's smaller than a grain of sand." *ExtremeTech*. Retrieved from: http://www.extremetech.com/extreme/155771-smart-dust-a-complete-computer-thats-smaller-than-a-grain-of-sand; accessed: Jan. 30, 2014.

The Cambridge Declaration on Consciousness: http://worldanimal.net/images/stories/documents/Cambridge-Declaration-on-Consciousness.pdf.

The difference between organism and animal: http://wikidiff.com/organism/animal

Thigpen, C. H. and Cleckley, H. M. *Three Faces of Eve* (New York: McGraw-Hill, 1957).

Thomas, R. L., Jiang, L., Orwoll, E. S., et al. (2020). "Vitamin D metabolites and the gut microbiome in older men." *Nature Communications*, 11(1), 1–10.

Thompson, S. V., Bailey, M. A., and Holscher, H. D. (2020). "Avocado Consumption Alters Gastrointestinal Bacteria Abundance and Microbial Metabolite Concentrations among Adults with Overweight or Obesity: A Randomized Controlled Trial." *Journal of Nutrition*.

Thurler, K. (2013). "Flatworms lose their heads but not their memories." Retrieved from: https://now.tufts.edu/news-releases/flatworms-lose-their-heads-not-their-memories.

Tilley, Sara, Neale, Chris, Patuano, Agnès, and Cinderby, Steve (2017). "Older People's Experiences of Mobility and Mood in an Urban Environment: A Mixed Methods Approach Using

Electroencephalography (EEG) and Interviews." *International Journal of Environmental Research and Public Health* 14(2), 151. Read more at https://medicalxpress.com/news/2017-04-green-spaces-good-grey.html#jCp.

Tirziu, D., Giordano, F. J., and Simons, M. (2010). "Cell communications in the heart." *Circulation*, 122(9), 928–937.

Tobi, Elmar W., Slieker, Roderick C., Xu, Kate M., et al. (2018). "DNA methylation as a mediator of the association between prenatal adversity and risk factors for metabolic disease in adulthood." *Science Advances*, 4(1).

Treffert, Darald. "Extraordinary People: Understanding the Savant Syndrome." (Lincoln, Neb.: iUniverse Inc, 2006).

Treffert, Darold (2015). "Genetic Memory: How We Know Things We Never Learned." *Scientific American*.

Trettenbrein, P. (2016). "The Demise of the Synapse As the Locus of Memory: A Looming Paradigm Shift?" *Frontiers in Systems Neuroscience,* 10. doi: 10.3389/fnsys.2016.00088.

Trewavas, A. (2002). "Plant intelligence: Mindless mastery." *Nature*, 415(6874), 841–841.

Trewavas, A. (a2005). "Green plants as intelligent organisms." *Trends in Plant Science*, 10(9), 413–419.

Trewavas, A. (b2005). "Plant Intelligence." *Naturwissenschaften*, 92(9), 401–413.

Tronick, E. *The Neurobehavioral and Social-Emotional Development of Infants and Children*. (New York: W. W. Norton, 2007).

Truman, J. W. (1990). "Metamorphosis of the central nervous system of Drosophila." *Journal of Neurobiology*, 21, 1072–1084.

Tsankova, N., Renthal, W., Kumar, A., and Nestler, E. J. (2007)." Epigenetic regulation in psychiatric disorders." *Nature Reviews Neuroscience*, 8, 355–367.

Tseng, A. and Levin, M. (2013). "Cracking the bioelectric code: Probing endogenous ionic controls of pattern formation." *Communicative & Integrative Biology*, 6(1), 1–8.

Tsuda S., Aono, M., and Gunji, Y. P. (2004). "Robust and emergent Physarum logical-computing." *BioSystems*, 73(1), 45–55.

Tully, T., Cambiazo, V., and Kruse, L. (1994). "Memory through metamorphosis in normal and mutant Drosophila." *Journal of Neuroscience*, 14, 68–74.

Tuominen, J., Kallio, S., Kaasinen, V., and Railo, H. (2021). "Segregated brain state during hypnosis." *Neuroscience of Consciousness*, 2021(1), niab002.

Turin, Luca. *The Secret of Scent: Adventures in Perfume and the Science of Smell*. (New York: Ecco, 2006).

Turner, C. H., Robling, A. G., Duncan, R. L., Burr, D. B. (2002). "Do bone cells behave like a neuronal network?" *Calcified Tissue International*, 70, 435-42.

Turner, S. G. and Hooker, K. (2020). "Are Thoughts About the Future Associated With Perceptions in the Present?: Optimism, Possible Selves, and Self-Perceptions of Aging." *The International Journal of Aging and Human Development*, 0091415020981883.

Tuszynski, Jack A. *The Emerging Physics of Consciousness*. (Berlin: Springer-Verlag, 2006).

Uchino, B. N., Cacioppo, J. T., and Kiecolt-Glaser, J. K. (1996). "The relationship between social support and physiological processes: A review with emphasis on underlying mechanisms and implications for health." *Psychological Bulletin*, 119(3), 488–531. doi:10.1037/0033-2909.119.3.488.

U.S. Surgeon General, U.S. Dept of Health & Human Services, (2017). The Surgeon General's Call to Action to Support Breastfeeding.

Umesono, Y., Tasaki, J., Nishimura, K., Inoue, T., and Agata, K. (2011). "Regeneration in an evolutionarily primitive brain—The planarian Dugesia japonica model." *European Journal of Neuroscience*, 34, 863–869.

University of Cambridge (2013). *Research News,* Molly Fox quoted.

Upledger, John E. *Your Inner Physician and You: Craniosacral Therapy and Somatoemotional Release* (Berkeley, Calif.: Atlantic Books, 1997).

UT Southwestern Medical Center (2017). "Muscle, not brain, may hold answers to some sleep disorders." *ScienceDaily*, www.sciencedaily.com/releases/2017/08/170803145629.htm.

Vahdat, Shahabeddin, Lungu, Doyon, Ovidiu, Julien, et al. (2015). "Simultaneous Brain-Cervical Cord fMRI Reveals Intrinsic Spinal Cord Plasticity during Motor Sequence Learning." *PLoS Biology*. http://dx.doi.org/10.1371/journal.pbio.1002186.

Vaidyanathan, G. (2017). "Science and Culture: Could a bacterium successfully shepherd a message through the apocalypse?" *Proceedings of the National Academy of Sciences*, 114(9), 2094–2095.

van der Kolk, B. A. *Psychological Trauma*. (Washington: American Psychiatric Press, 1987).

van der Kolk, B. A. (1994). "The body keeps the score: Memory and the evolving psychobiology of posttraumatic stress." *Harvard Review of Psychiatry*, 1(5), 253–265.

Van der Molen, M. W., Somsen, R.J.M., and Orlebeke, J. F. (1985). "The rhythm of the heart beat in information processing." *Advances in Psychophysiology*, 1, 1-88.

van der Windt, Gerritje J. W., Everts, Bart, Pearce, Erika L., et al. (2012). "Mitochondrial Respiratory Capacity Is a Critical Regulator of CD8+T Cell Memory Development." *Immunity*, 36(1), 68–78.

Van Etten, James L., Lane, Leslie C., and Dunigan, David D. (2010). "DNA Viruses: The Really Big Ones (Giruses)." *Annual Review of Microbiology*, 64, 83–99.

Vandvik, P. O., Wilhelmsen, I., Ihlebaek, C., and Farup, P. G. (2004). "Comorbidity of irritable bowel syndrome in general practice: a striking feature with clinical implications." *Alimentary Pharmacology & Therapeutics*, 20(10), 1195–1203.

Vattay, G., Kauffman, S., and Niiranen, S. (2014). "Quantum biology on the edge of quantum chaos." *PloS One*, 9(3), e89017.

Verdino, J. (2017). "The third tier in treatment: Attending to the growing connection between gut health and emotional well-being." *Health Psychology Open*, 4(2), 2055102917724335.

Verner, G., Epel, E., and Entringer, S. (2021). "Maternal psychological resilience during pregnancy and newborn telomere length: A prospective study." *American Journal of Psychiatry*, 178, 2.

Verny, Thomas R. *Pre-and-Peri-Natal Psychology: An Introduction*. (New York: Human Sciences Press, 1987).

Verny, Thomas R. and Kelly, John. *The Secret Life of the Unborn Child*. (New York: Summit Books, 1981).

Verny, Thomas R. and Weintraub, Pamela. *Pre-Parenting, Nurturing Your Child from Conception*. (New York: Simon & Schuster, 2002).

Villemure, J. G. and Rasmussen, T. H. (1993). "Functional hemispherectomy in children." *Neuropediatrics*, 24(1), 53–55.

Vining, E. P., Freeman, J. M., Pillas, D. J., Uematsu, S., and Zuckerberg, A. (1997). "Why would you remove half a brain? The outcome of 58 children after hemispherectomy—the Johns Hopkins experience: 1968 to 1996." *Pediatrics*, 100(2), 163–171.

Virophages: In *Wikipedia*. Retrieved from https://en.wikipedia.org/wiki/Virophage.

Volbrecht, M. M., Lemery-Chafant, K., Goldsmith, H. H., et al. (2007). "Examining the familial link between positive affect and empathy development in the second year." *Journal of Genetic Psychology*, 168, 105–129.

Volk, Steve (2018). "Down The Quantum Rabbit Hole." *Discover.*

Volkan, Vamik D. (1998). "Transgenerational Transmissions and 'Chosen Trauma'." Opening Address, XIII International Congress. International Association of Group Psychotherapy. http://www.vamikvolkan.com/Transgenerational-Transmissions-and-Chosen-Traumas.php.

Wagner, A. D. and Davachi, L. (2001). "Cognitive neuroscience: Forgetting of things past." *Current Biology*, 11, R964–967.

Wagner, M., Helmer, C., Samieri, C., et al. (2018). "Evaluation of the concurrent trajectories of cardiometabolic risk factors in the 14 years before dementia." *JAMA Psychiatry*, 75(10), 1033–1042.

Wakeup-world.com (2012). Retrieved from https://wakeup-world.com/2012/02/29/hearts-have-their-own-brain-and-consciousness/wuw_paginate/disabled/.

Walker, D., Greenwood, C., Hart, B., and Carta, J. (1994). "Prediction of school outcomes based on early language production and socioeconomic factors." *Child Development*, 65, 606–621.

Wang, H., Duclot, F., Liu, Y., Wang, Z. and Kabbaj, M. (2013). "Histone deacetylase inhibitors facilitate partner preference formation in female prairie voles." *Nature Neuroscience*, doi: 10.1038/nn.3420.

Ward, I. D., Zucchi, F. C., Robbins, J. C., Falkenberg, E. A., Olson, D. M., Benzies, K., and Metz, G. A. (2013). "Transgenerational programming of maternal behaviour by prenatal stress." *BMC Pregnancy and Childbirth*, 13(1), 1.

Warre-Cornish, K., Perfect, L., Srivastav, Deepak P., McAlonan, G., et al. (2020). "Interferon-γ signaling in human iPSC-derived neurons recapitulates neurodevelopmental disorder phenotypes." *Science Advances*, 6(34), eaay9506.

Waters, E., Merrick, S., Treboux, D., Crowell, J. and Albersheim, L. (2000), "Attachment Security in Infancy and Early Adulthood: A Twenty-Year Longitudinal Study." *Child Development*, 71, 684–689. doi:10.1111/1467-8624.00176.

Weaver, I.C.G., Cervoni, N., Champagne, F. A., Meaney, M. J., et al. (2004). "Epigenetic programming by maternal behavior." *Nature Neuroscience*, 7, 847–854.

Weaver, Kathryn, Campbell, Richard, Mermelstein, Robin, and Wakschlag, Lauren (2007). "Pregnancy Smoking in Context: The Influence of Multiple Levels of Stress." *Oxford Journals, Nicotine & Tobacco Research*, 10(6), 1065–1073.

Webre, D. J., Wolanin, P. M., and Stock, J. B. (2003). "Primer: bacterial chemotaxis." *Current Biology*, 13, R47–R49.

Westermann, J., Lange, T., Textor, J., and Born, J. (2015). "System consolidation during sleep–a common principle underlying psychological and immunological memory formation." *Trends in Neurosciences*, 38(10), 585–597.

Whitehead, A. N. *Process and Reality*. (New York: MacMillan, 1929).

Whitehead, A. N. *Adventure of Ideas*. (London: MacMillan, 1933).

Whiting, J. G., Jones, J., Bull, L., Levin, M., and Adamatzky, A. (2016). "Towards a Physarum learning chip." *Scientific Reports*, 6.

WHO, Nutrition, Exclusive breastfeeding 2011-01-15; and UNICEF (n.d.). "Skin-to-skin contact." http://www.unicef.org.uk/BabyFriendly/News-and-Research/Research/Skin-to-skin-contact/.

Wildschutte, Julia Halo, Williams, Zachary H., Coffin, John M., et al. (2016). "Discovery of unfixed endogenous retrovirus insertions in diverse human populations." *Proceedings of the National Academy of Sciences*, 113(16), E2326-E2334.

Wittmann, Marc. *Felt Time: The Psychology of How We Perceive Time*. (Boston: MIT Press, 2016).

Wittmann, Marc. *Altered States of Consciousness: Experiences Out of Time and Self.* (Boston: MIT Press, 2018).

Wixted, J. T., Squire, L. R., Jang, Y., Papesh, M. H., Goldinger, S. D., Kuhn, J. R., and Steinmetz, P. N. (2014). "Sparse and distributed coding of episodic memory in neurons of the human hippocampus." *Proceedings of the National Academy of Sciences*, 111(26), 9621–9626. 1.

Wojtowicz, J. M. (2011). "Adult neurogenesis. From circuits to models." *Behavioral Brain Research*. [Epub ahead of print]. 10.1016/j.bbr.2011.08.013.

Wong, G. C., Antani, J. D., Bassler, Bonnie, Dunkel, J., et al. (2021). "Roadmap on emerging concepts in the physical biology of bacterial biofilms: from surface sensing to community formation." *Physical Biology*.

Wolf, D. M., Fontaine-Bodin, L., Bischofs, I., Price, G., Keasling, J., and Arkin, A. P. (2008). "Memory in microbes: Quantifying history-dependent behavior in a bacterium." *PLoS one*, 3(2), e1700.

Wu, C. L., Shih, M. F., Chiang, A. S., et al. (2011). "Heterotypic Gap Junctions between Two Neurons in the Drosophila Brain Are Critical for Memory." *Current Biology*, 21(10), 848–854.

Yan, Z., Lambert, N. C., Guthrie, K. A., Porter, A. J., Nelson, J. L., et al. (2005). "Male microchimerism in women without sons: Quantitative assessment and correlation with pregnancy history." *American Journal of Medicine*, 118(8), 899–906.

Yan, Jian, Enge, Martin, Taipale, Minna, Taipale, Jussi, et al. (2013). "Transcription Factor Binding in Human Cells Occurs in Dense Clusters Formed around Cohesin Anchor Sites." *Cell*, 154(4), 801.

Yang, Fu-liang in Savage, Sam (2010). "Scientists Create World's Smallest Microchip." RedOrbit .com, http://www.redorbit.com/news/technology/1966110 scientists_create_worlds_smallest _microchip/#7s4r0ZiF0uvXwP5B.99.

Yang, S., Platt, R. W., and Kramer, M. S. (2010). "Variation in child cognitive ability by week of gestation among healthy term births." *American Journal of Epidemiology*, 171(4), 399–406.

Yao, Youli, Robinson, Alexandra M., Metz, Gerlinde, A. S., et al. (2014). "Ancestral exposure to stress epigenetically programs preterm birth risk and adverse maternal and newborn outcomes." *BMC Medicine*, 12(1), 121. doi: 10.1186/s12916-014-0121-6.

Yehuda, R., Daskalakis, N. P., Lehrner, A., Desarnaud, F., Bader, H. N., Makotkine, I., et al. (2014). "Influences of maternal and paternal PTSD on epigenetic regulation of the glucocorticoid receptor gene in Holocaust survivor offspring." *American Journal of Psychiatry*, 171(8), 872–8010.1176/appi.ajp.2014.13121571.

Yehuda, R., et al. (2005). "Transgenerational Effects of Posttraumatic Stress Disorder in Babies of Mothers Exposed to the World Trade Center Attacks during Pregnancy." *Journal of Clinical Endocrinology & Metabolism*. doi: 10.1210/jc.2005-0550.

Yehuda, R., Halligan, S. L., and Grossman, R. (2001). "Childhood trauma and risk for PTSD: Relationship to intergenerational effects of trauma, parental PTSD, and cortisol excretion." *Development and Psychopathology*, 13(3), 733–753.

Yehuda, R. and Lehrner, A. (2018). "Intergenerational transmission of trauma effects: Putative role of epigenetic mechanisms." *World Psychiatry*, 17(3), 243–257.

Yehuda, Rachel, Daskalakis, Nikolaos P., Desarnaud, Frank, et al. (2013). "Epigenetic biomarkers as predictors and correlates of symptom improvement following psychotherapy in combat veterans with PTSD." *Frontiers in Psychiatry*. doi: 10.3389/fpsyt.2013.00118.

Yehuda, Rachel, Daskalakis, Nikolaos P., Binder, Elisabeth B., et al. (2016). "Holocaust Exposure Induced Intergenerational Effects on FKBP5 Methylation." *Biological Psychiatry*, 80(5), 372. doi: 10.1016/j.biopsych.2015.08.005.

Yeragani, V. K., Nadella, R., Hinze, B., Yeragani, S., and Jampala, V. C. (2000). "Nonlinear measures of heart period variability: Decreased measures of symbolic dynamics in patients with panic disorder." *Depression and Anxiety*, 12, 67–77.

Yeragani, V. K., Rao, K. A., Smitha, M. R., Pohl, R. B., Balon, R., and Srinivasan, K. (2002). "Diminished chaos of heart rate time series in patients with major depression." *Biological Psychiatry*, 51, 733–744.

Yoney, A., Etoc, F., Brivanlou, A. H., et al. (2018). "WNT signaling memory is required for ACTIVIN to function as a morphogen in human gastruloids." *eLife*, 7, e38279.

Younge, N., McCann, Seed, P. C., et al. (2019). "Fetal exposure to the maternal microbiota in humans and mice." *JCI Insight*.

Yu, C., Guo, H., Cui, K., Li, X., Ye, Y. N., Kurokawa, T., and Gong, J. P. (2020). "Hydrogels as dynamic memory with forgetting ability." *Proceedings of the National Academy of Sciences*.

Zammit, Stanley, et al. (2009). "Maternal tobacco, cannabis and alcohol use during pregnancy and risk of adolescent psychotic symptoms in offspring." *The British Journal of Psychiatry*, 195, 294–300.

Zamroziewicz, Marta K., Paul, Erick J., Zwilling, Chris E., and Barbey, Aron K. (2017). "Determinants of fluid intelligence in healthy aging: Omega-3 polyunsaturated fatty acid status and frontoparietal cortex structure." *Nutritional Neuroscience,* 1–10.

Zarnitsyna, V. I., Huang, J., Zhang, F., Chien, Y. H., Leckband, D., and Zhu, C. (2007). "Memory in receptor-ligand-mediated cell adhesion." *Proceedings of the National Academy of Sciences,* 104(46), 18037–18042. http://www.pnas.org/content/104/46/18037.short.

Zayed, Amro and Robinson, Gene E. (2012). "Understanding the Relationship Between Brain Gene Expression and Social Behavior: Lessons from the Honey Bee." *Annual Review of Genetics,* 46, 591–615.

Zenger, M., Glaesmer, H., Höckel, M., and Hinz, A. (2011). "Pessimism predicts anxiety, depression and quality of life in female cancer patients." *Japanese Journal of Clinical Oncology,* 41(1), 87–94.

Zhang, W. B., Zhao, Y., and Kjell, F. (2013). "Understanding propagated sensation along meridians by volume transmission in peripheral tissue." *Chinese Journal of Integrative Medicine,* 19(5), 330–339.

Zhu, L., Aono, M., Kim, S. J., and Hara, M. (2013). "Amoeba-based computing for traveling salesman problem: Long-term correlations between spatially separated individual cells of Physarum polycephalum." *Biosystems,* 112(1), 1–10.

Zimmer, Carl (2011). "Can Answers to Evolution Be Found in Slime?" *New York Times.*

Zoghi, M. (2004). "Cardiac memory: Do the heart and the brain remember the same?" *Journal of Interventional Cardiac Electrophysiology,* 11, 177–178.

INDEX